Advances in Fossil Energy R&D 2018

Topical Conference at the 2018 AIChE Annual Meeting

Pittsburgh, Pennsylvania, USA
28 October - 2 November 2018

ISBN: 978-1-5108-7642-2

Printed from e-media with permission by:

Curran Associates, Inc.
57 Morehouse Lane
Red Hook, NY 12571

Some format issues inherent in the e-media version may also appear in this print version.

Printed by Curran Associates, Inc. (2019)

For permission requests, please contact AIChE
at the address below.

AIChE
120 Wall Street, FL 23
New York, NY 10005-4020

Phone: (800) 242-4363
Fax: (203) 775-5177

www.aiche.org

Additional copies of this publication are available from:

Curran Associates, Inc.
57 Morehouse Lane
Red Hook, NY 12571 USA
Phone: 845-758-0400
Fax: 845-758-2633
Email: curran@proceedings.com
Web: www.proceedings.com

Advances in Fossil Energy R&D 2018

Topical Conference at the 2018 AIChE Annual Meeting

Pittsburgh, Pennsylvania, USA
28 October - 2 November 2018

AIChE
The Global Home of Chemical Engineers

TABLE OF CONTENTS

(11a) Fabrication and Field Testing of Spiral-Wound Membrane Modules for CO_2 Capture from Flue Gas 1
Witopo Salim, Varun Vakharia, Yuanxin Chen, Dongzhu Wu, Yang Han, W. S. Winston Ho

(11b) Low Viscosity Water-Lean Diamine Solvents for Carbon Dioxide Capture 2
Phillip K. Koech, Deepika Malhotra, Manh Nguyen, David J. Heldebrant, Andy J. Zwoster, Vassiliki-Alexandra Glezakou, Feng Zheng, Roger Rousseau

(11c) High Swing Capacity MIL-101(Cr) Fiber Sorbents for Sub-Ambient CO_2 Capture Via RCPSA 3
Stephen J. A. Dewitt, Rohan Awati, Matthew Realff, David S. Sholl, Ryan Lively

(11d) Carbon Dioxide Capture By an Electrochemically-Mediated Amine Regeneration 4
Mohammad Rahimi, Miao Wang, Subrahmaniam Hariharan, Michael Massen-Hane, T. Alan Hatton

(11e) Layer-By-Layer Functional Thin Film Coatings for Enhanced Light Gas Separations 5
Benjamin Wilhite, Jaime C. Grunlan

(11f) Facilitated Transport Membranes for Hydrogen Purification from Coal-Derived Syngas 6
Yang Han, W. S. Winston Ho

(11g) Novel Membranes for CO_2 concentration and Carbon Capture 7
Ning Shangguan, Kenneth J. Pennisi

(11h) Process Development of Novel Sub-Ambient Membrane Processes for CO_2 Capture 8
Jin-Kuk Kim, Sunghoon Lee, Seokwon Yun

(58a) U.S. DOE National Energy Technology Laboratory: Carbon Capture R&D Program's New Direction for Operational Versatility 9
J. D. Figueroa, Lynn Brickett, Kanwal Mahajan

(58b) Highlights of Collaboration between the U.S.DOE Carbon Capture Simulation for Industry Impact Program and the Discovery of Carbon Capture Substances and Systems Initiative 10
Michael S. Matuszewski, Benjamin P. Omell, David C. Miller, Debangsu Bhattacharyya, Rebecca Siegelman, Jeffrey R. Long, Charles J. Freeman, Zhijie Xu, Joshuah Stolaroff

(58c) Process Modeling and Experimental Studies of a Diamine-Appended Metal-Organic Framework for CO_2 Capture 11
Ryan Hughes, Goutham Kotamreddy, Debangsu Bhattacharyya, Michael S. Matuszewski, Rebecca Siegelman, Jeffrey R. Long

(58d) Analysis of Different Solvent Performance in UKy-CAER's 0.7 MWe CO_2 Capture Pilot Plant 12
Reynolds A. Frimpong, Heather Nikolic, Jonathan V. Pelgen, Kunlei Liu

(58e) In-Situ Gasification Chemical Looping Combustion vs. Chemical Looping with Oxygen Uncoupling: Exergy Comparison for Power Generation with CO_2 Capture 13
Yitao Zhang, Andrew Tong

(58f) The Development of Machine Learning, Group Contribution and Molecular Modeling Approach to Screen Physical Solvents for Gas Separation 14
Wei Shi, Megan Macala, Robert L. Thompson, Surya Tiwari, Kevin P. Resnik, Nicholas Siefert, David Hopkinson

(58g) Towards the Development of a Solvent Screening Tool for CO_2 Capture Using Molecular Thermodynamics 15
Luis M. C. Pereira, F. Llovell, Ismail I. Alkhatib, Lourdes F. Vega

(58h) Systematic Design of Phase-Change Solvents for Post-Combustion CO_2 Capture Based on Advanced Thermodynamics and Holistic Sustainability Assessment 16
Athanasios I. Papadopoulos, Gulnara Shavalieva, Felipe Perdomo-Hurtado, Panos Seferlis, Stavros Papadokonstantakis, Claire S. Adjiman, Amparo Galindo, George Jackson

(120b) 2018 Outlook for Energy: A View to 2040 17
Theodore J. Wojnar Jr.

(120c) Energy Decarbonisation Scenarios 18
Kamel Ben Naceur

(120a) Fundamental Research Needs to Advance Energy Technologies 19
Bruce Garrett

(147a) An Overview of Recent Efforts Under the Department of Energy's Carbon Storage Program: Moving CCS Towards Commercialization 20
Mary Sullivan, Erik Albenze, Kanwal Mahajan, Traci Rodosta

(147c) Chemical Impacts of CO_2 Intrusion into Heterogeneous Caprock 21
Ting Xiao, Brian McPherson, Nathan Moodie, Trevor Irons, Wei Jia, Richard Esser

(147d) Implementing and Validating Active Reservoir and Brine Management Strategies for the Enhancement of the Geologic Storage of Carbon Dioxide: An Update 22
David Nakles, John A. Hamling, Ryan J. Klapperich

(147e) Evolution of Transport and Mechanical Properties of Mt Simon Sandstones Due to Interaction with Brine/CO_2 23
Zhuofan Shi, Lin Sun, Kristian Jessen, Theodore Tsotsis

(147f) Optimization of CO2-Enhanced Oil Recovery with CO2 Storage in a Mature Oil Field 24
William Ampomah, Robert Balch, Robert Will, Reid Grigg, Martha Cather

(147g) How Much Oil and CO_2? at What Cost? Bounding the Quantities of Produced Oil and Stored CO_2 from CO_2 EOR in Conventional Oil Fields in the United States 25
David Morgan, Donald Remson, Tim Grant

(187b) Ultra-Deep Desulfurization of Low-Sulfur Gasoline By Selective Adsorption of Trace Mercaptans over Supported Metal Oxides 26
Cuiting Yang, Guang Miao, Zhong Li, Jing Xiao

(187c) Catalytic Adsorptive Desulfurization (CADS) of Diesel Using Industrial-Grade MCM-41 27
Xiong Dai, Guang Miao, Zhong Li, Jing Xiao

(187f) Evolution of CO_2 Storage Capacity Associated with Geochemical Reactions in Subsurface 28
Wei Jia, Ting Xiao, William Ampomah, Nathan Moodie, Brian McPherson

(187g) Quantitative Analysis of the Influence of Capillary Pressure on Geologic Carbon Storage Forecasts. Case Study: CO_2-EOR in the Anadarko Basin, Texas 29
Nathan Moodie, William Ampomah, Wei Jia, Jason Heath, Brian McPherson

(187h) Measurement and Calibration of Self-Sealing Rate of Fractures in Geological CO_2 Storage: Case Study of a Natural Analog 30
Vivek Patil, Brian McPherson, Edward Trujillo, Hyukmin Kweon

(187q) New Data and Models to Avoid Cryogenic Solids Formation in LNG Production 31
Arman Siahvashi, Saif Zs. Al-Ghafri, Brendan F. Graham, Eric F. May

(187i) An Innovative Technology for the Production of Value Added Chemicals Using Fischer Tropsch Synthesis 32
Syed Ali Zeeshan Gardezi

(187j) Strongly Coupled Co@CoO_x Nanoparticles and Layered Perovskite As a Highly Stable and Efficient Cathode for Solid Oxide Electrolysis Cells 33
Yongdan Li

(187k) Bimetallic Pd-Cu Catalysts for CO_2 Hydrogenation to Methanol 34
Xiao Jiang, Xiaowa Nie, Xinwen Guo, Krista S. Walton, Chunshan Song

(187l) Energy Analysis of Non-Aqueous Solvents (NASs) for CO_2 Capture Process 35
Aravind V. Rayer, Paul Mobley, Vijay Gupta, Jak Tanthana, Mustapha Soukri, Marty Lail, S. James Zhou

(187m) Microwave Assisted Lignin Depolymerization Using Deep Eutectic Solvents 36
Pranjali Muley, Dorin Boldor, Jian Shi, Bert C. Lynn

(187n) Enhanced Gasification Reactor Designs for Maximizing Gas-Particle Interaction 37
Quang Truong, Srujan Rokkam, Matt Flannery

(187o) Modeling the Decomposition Kinetics of the Gas Hydrates in Porous Medium 38
Avinash V. Palodkar, Amiya Kumar Jana

(187p) Modeling Phase Equilibrium with Wong-Sandler Mixing Rule for Ternary $CO2/H_2/C_2H_6$ Hydrates 39
Niraj Thakre, Amiya Kumar Jana

(235i) The U.S. Department of Energy's R&D Program for Carbon Use and Reuse 40
Lynn Brickett, John Litynski

(235a) Enabling a Solid-State Carbon Dioxide Distribution Network 41
Kevin Blinn, Daniel Kopp, Jun Wang, Richard E. Riman

(235b) Plasma and Fluidic Oscillation Assisted Electrolysis of CO_2 42
Rachael H. Rothman, Ann V. Call, Tom Butterworth, Thomas Holmes, Lik Hang Hugo Tse, Pratik Desai, William B. Zimmerman

(235c) Treatment and Extraction of Metals from Electronic Wastes Using a Novel Solvent Containing Supercritical CO_2 43
Emily Hsu, Ah-Hyung Alissa Park, Alan C. West, Katayun Barmak

(235e) A CO_2 Utilization Approach Towards the Synthesis of Terephthalic Acid 44
Samuel Thompson, Marty Lail, Qinghe Zheng, Thomas Gohndrone

(235f) Catalytic CO_2 Hydrogenation to C_2^+ Hydrocarbons 45
Wenjia Wang, Xiaoxing Wang, Xiao Jiang, Chunshan Song

(235g) Effect of CO_2 Source and Concentration on the Catalyst Performance and Economics of Ethane Oxidative Dehydrogenation to Ethylene 46
Jadid E. Samad, Amit Goyal

(235h) Utilization of CO_2 Towards Solar Fuel Production Via Ferrite Based Redox Reactions 47
Rahul Bhosale, Gorakshnath Takalkar, Aliya Banu, Fares AlMomani, Anand Kumar, Majeda Khraisheh
(274a) Efficient Selection of Conventional and Phase-Change CO_2 Capture Solvents Based on Nominal and Off-Design Process Operation 57
Theodoros Zarogiannis, Athanasios I. Papadopoulos, Ioannis Tsivintzelis, Panos Seferlis
(274b) Modular CO_2 Capture from Distributed Oilfield Engines 59
Joseph Moate, Tony Neumayer, Phil Dipietro, Charles Womble
(274c) Application of Dynamic Reduced-Order Modeling and Advanced Process Control on UKy-CAER CO_2 Capture Pilot Plant Using CCSI Tools 60
Priyadarshi Mahapatra, Jinliang Ma, Benjamin P. Omell, Michael S. Matuszewski, Jonathan V. Pelgen, Kunlei Liu
(274d) An Optimization-Based Methodology for the Reduction of Gas Flaring in Shale Oil Production 61
A. J. C. Vergara, Natalie J. Pekney
(274e) Techno-Economic System Analysis for SOFC/GT Hybrid System Accounting for Degradation Effects 62
Haoxiang Lai, Thomas A. Adams II
(274f) Coal Drying, Mercury Removal, and Stabilization Using Heat from a Nuscale Small Modular Reactor 63
Ying Wang, William C. Schaffers, David A. Bell, Jong Suk Kim, Richard Boardman
(274g) Optimal Design of Gas-Fired Moving-Bed Chemical Looping Combustion Systems 64
Anca Ostace, Chinedu O. Okoli, Andrew Lee, Anthony P. Burgard, Debangsu Bhattacharyya, David C. Miller
(311a) The Impact of Shale Gas and Oil on the Chemical Industry 65
Jeffrey J. Siirola
(311b) Sustainable Energy and Chemicals: Past, Present, and Future 66
Joseph B. Powell
(311c) Disruptions: What the Future May Hold 67
Scott F. Mitchell
(311d) Geopolitical Factors Influencing the Evolution of the Chemical Industry 68
David West
(311e) Agility & Resilience: How to Maintain Career Competitiveness in the Changing Chemical Industry 69
Antonis Papadourakis
(322a) Catalytic Aromatization of Bio-Derived Liquid Under Methane Environment 70
Aiguo Wang, Danielle Austin, Hua Song
(322b) Wave Liquefaction™ Processing of Carbon Materials for the Production of Value-Added Chemicals and Feedstocks: Spectroscopic Diagnostics and Material Characterization 71
Randy Vander Wal, Arupananda Sengupta, Madhu Singh, Kurt Zeller, George Skoptsov
(322c) Hydrocarbon Chain Growth Via a Nonthermal Electrical Plasma Microreactor 74
Ian Reddick, Goran Jovanovic, Alexandre Yokochi, Nick Auyeung, Matthew Young Coblyn, Yu Miao, Omar Mohamed, Adam Shareghi, Andrew Traverso, Anthony Pyka
(322d) Combined Experimental and Density Functional Theory Studies on the Modified $SrTiO_3$ Catalysts for Oxidative Coupling of Methane 75
Seo Yeon Lim, Jae-Wook Choi, Dong Jin Suh, Kwang Ho Song, Hyung Chul Ham, Jeong-Myeong Ha
(322e) Process Technology Assessment for Mega Scale Projects 76
Yizu Zhu, Mohammad Shafiei
(370b) Catalyst Development for Natural Gas Monetization 77
Daniyal Kiani, Sagar Sourav, Lohit Sharma, Israel E. Wachs, Jonas Baltrusaitis
(370c) Catalytic Wall Reactor for Non-Oxidative Methane Conversion 78
Su Cheun Oh, Dongxia Liu
(370d) Exploring Strategies to Improve Yields of Oxidative Coupling of Methane in a Chemical Looping System 79
Deven Baser, Zhuo Cheng, Sourabh Nadgouda, Lang Qin, Liang-Shih Fan
(370e) Metal-Mediated Transient Hydrogen Scavenging for Enhanced Aromatics Yield during Non-Oxidative Methane Aromatization on Mo/H-ZSM-5 Catalysts 80
Anurag Kumar, Aditya Bhan
(370f) Promotional Effect of Cr in Mo_2C_A catalyst Supported on Sulfated Zirconia for Methane Dehydroaromatization 81
Ashraf Abedin, Swarom Kanitkar, Srikar Bhattar, Jim Spivey
(370g) Engineering Fe-HZSM-5 for Methane Dehydroaromatization 82
Yifan Deng, Yu-Chieh Cheng, Yahui Yang, Yungchieh Lai, G. Veser

(370h) Magnesium Vanadate Catalyzed Oxidative Dehydrogenation of Ethane to Ethylene Using CO_2 As a Soft Oxidant 83
Chinmoy Baroi, Harry W. Rollins, Rebecca Fushimi
(370i) Chemical Looping for the Oxidative Cracking of Shale Condensates 84
Luke Neal, Vasudev Pralhad Haribal, Fanxing Li
(430a) The U.S. Department of Energy's Fossil Energy Research and Development Priorities 85
Steven Winberg
(430b) Low-Carbon Solution to Xtl&c and Its Progress of Key Technologies 86
Yuhan Sun
(430c) The Global Energy Challenge: What's Ahead 87
Robert C. Armstrong
(430d) Advanced Modeling and Optimization for Future Generation Energy Systems 88
Lorenz T. Biegler
(439a) High Purity Hydrogen Production in a 10 kW_{th} Fixed Bed RESC Prototype System 89
Sebastian Bock, Robert Zacharias, Viktor Hacker
(439b) High Purity Syngas and Hydrogen Coproduction from Natural Gas Using Cu-Fe Based Metal Oxides in a Chemical Looping System 90
Sourabh Nadgouda, Mengqing Guo, Liang-Shih Fan
(439c) Process System Analysis of a High-Pressure Chemical Looping Based Hydrogen Production System 91
Mandar Kathe, Frank Kong, Kate Clelland, Tyler Christeson, Andrew Tong, Liang-Shih Fan
(439d) Methane to Syngas By Chemical Looping Using FE-NI Oxygen Carriers: Reactor Design and Process Modeling 92
Hari C. Mantripragada, G. Veser
(439e) Dry Reforming of Methane Over Ni Based Lanthanum Zirconate Pyrochlore Catalysts: Deactivation Study 93
Srikar Bhattar, Swarom Kanitkar, Ashraf Abedin, Dushyant Shekhawat, Daniel J. Haynes, Jim Spivey
(439f) Pressure Dilution, a New Method to Prepare a Stable Ni/Fumed Silica Catalyst for the Dry Reforming of Methane to Produce Hydrogen 94
Eduardo E. Wolf, V. Danghyan, S. Calderon Novoa, A. Mukasyan
(439g) A Novel Carbon-Resistant Perovskite Catalyst for Hydrogen Production Using Methane Dry Reforming 95
Feraih Alenazey, Yousef Alyousef, Raja Al Otaibi, Adesina Adesoji A., Faisal Alotaibi, Bandar Alotaibi, Ghzzai Almutairi, Dai-Viet N. Vo
(514a) Water-Gas Shift Reactor for Fuel Cell Systems: Stable Operation for 5000 Hours 96
Joachim Pasel, Remzi Can Samsun, Andreas Tschauder, Ralf Peters, Detlef Stolten
(514b) Enhancements on Diesel Autothermal Reformer for Low-Temperature Solid Oxide Fuel Cell Integration 97
Minseok Bae, Hyungjun Cheon, Jiwoo Oh, Joongmyeon Bae, Sai P. Katikaneni
(514c) Rapid Production of High-Purity Hydrogen Fuel from Liquid Hydrocarbon Fossil Fuels 98
Xiangyu Jie, Peter P. Edwards, Tiancun Xiao
(514d) Hydrogen Production from Liquid Hydrocarbons with Process Intensification - a Case Study 99
Jaemyung Lee, Jae Young Yoo, Joongmyeon Bae, Aadesh X. Harale, Sai P. Katikaneni
(514e) The Effect of Catalytic Metal on the Activity of Lanthanum Zirconate Pyrochlore Under Low S/C Methane Reforming Conditions 100
Daniel J. Haynes, Dushyant Shekhawat, David Berry, Jim Spivey
(514f) Hydrogen Production through Diesel Steam Reforming: Process Intensification Using Structured Thermo-Neutral Reforming Catalyst 101
Shakeel Ahmed, Sai P. Katikaneni, Aadesh X. Harale
(514g) High Temperature Catalyst Development and Testing for Low Cost and Efficient Solar Driven Sulfur-Based Hydrogen Production 102
Birendra Adhikari, Daniel M. Ginosar, Fahim Rahman, Weijian Diao, John Meynard M. Tengco, John Monnier, Claudio Corgnale
(514h) Technoeconomic Study of Advanced H_2 Production Technologies: Membrane-Supported H_2O Splitting, Thermochemical Redox H_2O Splitting and Fuel-Assisted H_2O Electrolysis 103
Xiao-Yu Wu, Ahmed F. Ghoniem
(570a) Microwave Chemistry 104
Dushyant Shekhawat, Mark W. Smith, Victor Abdelsayed, Christina Wildfire
(570b) Low-Temperature Microwave Plasma Conversion of Methane to Higher Hydrocarbons 105
George Skoptsov, Kurt Zeller, Randy Vander Wal
(570d) Catalyst Design for Microwave-Assisted Dry Reforming of Methane 106
Christopher Marin, Douglas R. Kauffman

(570e) Techno-Economic Analysis of Direct Non-Oxidative Conversion of Shale Gas Via Non-Thermal Microwave (MW) Plasma Catalysis 107
Chirag Mevawala, Jianli Hu, Debangsu Bhattacharyya
(570f) Novel Microwave Assisted Ammonia Synthesis By Methane and Nitrogen Under Atmospheric Pressure 108
Xinwei Bai, Sarojini Tiwari, Brandon Robinson, Jianli Hu
(570g) Spectroscopic and Microscopic Characterization of Nanographene Product By Microwave-Assisted Plasma-Mediated Methane Pyrolysis 109
Randy Vander Wal, Madhu Singh, Arupananda Sengupta, Kurt Zeller, George Skoptsov
(570h) Performance of Microwave-Assisted Catalysis for Production of Upgraded Pyrolysis Bio-Oil from Poplar Species in a Continuous-Flow System 112
Dorin Boldor, Cosmin Marculescu, Razvan N. State, Mariana Patrascu
(633a) Recovering Rare Earth Elements from Coal-Based Materials 113
Mary Anne Alvin
(633b) Transmission Electron Microscopy Observations of Rare Earth Minerals in the Fire Clay Coal, Kentucky, and in the Derived Combustion Ashes 114
James C. Hower
(633c) Selectivity and Recovery of Rare Earth Elements from Coal Ash Leachates Using Liquid Membrane-Based Separations 115
Ryan Smith, Ross Taggart, Mark R. Wiesner, Heileen Hsu-Kim
(633d) Effect of Calcium Halide Salt Addition on the Fate of Rare Earth Compounds during Coal Combustion Process 116
Ward A. Burgess, Murphy J. Keller, Elliot Roth, Bret H. Howard, Jonathan W. Lekse, Evan J. Granite
(633e) Examination of the Distribution and Form of the Rare Earth Elements in a Metalliferous Lignite Coal 117
Brittany Rew, Dan Laudal, Steve Benson, Nolan Theaker
(633f) Economic Extraction, Recovery and Upgrading of Rare Earth Elements from Coal-Based Resources 118
Michael Free
(633g) Hydrometallurgical Circuits for the Recovery of RARE Earth Elements from Coal Sources 119
Rick Honaker
(633h) Determination and Recovery of Rare Earths from Coal Combustion By-Products 120
Evan J. Granite, Ken Ladwig, Elliot Roth
(633i) Evaluation of Critical Trace Elements Including Rare Earth Elements in U.S. Coals 121
Ronghong Lin, Yee Soong, Evan J. Granite
(646a) Treatment Approaches for Produced Water Re-Use and Surface Discharge 122
Madison Wenzlick, Alexandra Hakala, Nicholas Siefert
(646b) Advanced Supercritical Water-Based Process Concepts for Treatment and Beneficial Reuse of Brine Generated By Oil/Gas Production 123
Jason Trembly, Chad Able, David Ogden
(646d) Permian & Anadarko Basin Produced Water Recycling: Keys to Success 124
J. Daniel Arthur, David Alleman
(655a) Catalytic Etherification of Glycerol to Glycerol Oligomers in the Presence of Alumina Supported Ca/Sr Mixed Oxides 125
Yi-Chen Shih, Bing-Hung Chen
(655b) The Role of Copper Stability in Selectively Condensation of Ethanol to Higher Alcohols 126
Mond Guo, Karthikeyan K. Ramasamy
(655c) Effects of Alloying Pd and Cu on Tandem Dehydrogenation-Aldol Condensation Reactions 127
Konstantinos A. Goulas, Yuying Song, Lars C. Grabow, Dean Toste
(655d) Lubricant Base Oils Production from Biomass 128
Sibao Liu, Basudeb Saha, Dionisios G. Vlachos
(655e) Selectivity Control During the One-Pot Conversion of Aliphatic Carboxylic Acids to Linear Olefins through Tandem Hydrogenation/Dehydration 129
Jher Hau Yeap, Bartosz Rozmyslowicz, Jeremy S. Luterbacher
(655f) Mechanism and Kinetics of Isobutene Production over Zirconia-Supported Zinc Oxides 130
Julie Rorrer, Alexis T. Bell, Dean Toste
(655g) Catalytic Upgrading of Sugar-Derived Polyols to Glycols in Absence of Externally Added Hydrogen 131
Bin Yin, Xin Jin, Guangyu Zhang, Hao Yan, Chaohe Yang
(655h) Selective C-C Bond Scission of Primary Alcohols Using Cerium Oxide-Supported Palladium Catalyst 132
Tomoo Mizugaki, Kodai Nitta, Takato Mitsudome, Koichiro Jitsukawa

(677a) New Technologies to Enhance the Production of Unconventional Oil and Natural Gas 133
Jared Ciferno
(677b) Characterization of Microfractures in Organic-RICH Shales and Tight Reservoir Rocks of the Bakken Formation By Integrated Microscopy Techniques 134
Alexander Azenkeng, Blaise Mibeck, Kurt Eylands, Shane Butler, Bethany Kurz
(677c) Remediation of Polyacrylamide-Induced Permeability Damage from Fracturing Using H_2O_2 Catalyzed By Free and Immobilized Peroxidase Enzyme 135
William J. R. Gilbert, Jyun Syung Tsau, Stephen J. Johnson, Jenn-Tai Liang, Aaron M. Scurto
(677d) Investigation of Scaling As a Means for Decreased Petroleum Production from the Utica/Point Pleasant Unconventional Play 136
Michael Spencer, Jason Trembly, Ravinder Garlapalli
(677e) CO_2-in-Mineral Oil Emulsions, CH_4-in-Mineral Oil Foams and N_2-in-Mineral Oil Foams Stabilized By Novel Oil-Soluble Surfactants As Waterless Hydraulic Fracturing Fluids 137
Shehab Alzobaidi, Gianfranco Rodriguez, Jason J. Lee, Congwen Lu, Chang Da, Justin Harris, Robert J. Perry, Keith Johnston, Robert Enick
(677f) Hydraulic Fracture Propagation in Unconventional Reservoirs Under the Influence of Natural Fracture Heterogeneities 138
Wei Fu, Alexei A. Savitski, Branko Damjanac, Andrew P. Bunger
(677g) Characterization of CO_2, Fluid, and Shale Via Feature Relocation Using Field-Emission Scanning Electron Microscopy, in Situ Infrared Spectroscopy, and Pore Size Analysis 139
Angela Goodman, Sean Sanguinito, Barbara Kutchko, Jeffery Culp, Sittichai Natesakhawat, Dustin Crandall
(695a) Gas Phase Catalytic Oxidation of Lignin to Produce Phenolic Compounds over Vanadia Catalysts 140
Matthew M. Yung, Calvin Mukarakate, Mark Nimlos, Michael B. Griffin, Seonah Kim, Eric C. D. Tan
(695b) Mechanistic Study of the Hydrogenolysis of Diaryl Ethers Catalyzed By Heterogeneous Metal Catalysts 141
Meng Wang, Oliver Gutierrez, Donald M. Camaioni, Johannes A. Lercher
(695c) Palladium-Iron Bimetallic Catalyst: High Activity and Stability for Aqueous Phase Hydrogenations 142
Yan Cheng, Hien N. Pham, Robert L. Johnson, Brent H. Shanks, Abhaya K. Datye
(695d) Hydrodeoxygenation of Guaiacol over Ni and Mo Nanoparticles Supported on SBA-15 and γ-Al_2O_3. 143
Thiago L. R. Hewer, Rubens W. S. Lima, Reinaldo Giudici, Martin Schmal, Rita M. B. Alves
(695e) Hydroprocessing of Biomass-Derived Oxygenates on Metal-Exchanged Zeolites Using Light Alkanes As the Source of Hydrogen 146
Dante Simonetti, Eric Lin
(695f) Flowthrough Reductive Catalytic Fractionation of Biomass 147
Eric Anderson, Michael Stone, Rui Katahira, Michelle Reed, Gregg T. Beckham, Yuriy Roman-Leshkov
(695h) Hydrodeoxygenation of Sorbitol to Monofunctional Fuel Precursors over Co/TiO_2 148
Nathaniel Eagan, Joseph P. Chada, Ashley Wittrig, J. Scott Buchanan, George W. Huber, James A. Dumesic
(713a) Numerical Simulation of Natural Fractures on EOR By Cyclic Gas Injection in Unconventional Reservoirs Using DFN Approach 149
Chongwei Xiao, Olatoyosi Obilade
(713b) Enhancing Hydraulic Fracturing Productivity Via Model-Based Feedback Control 162
Prashanth Siddhamshetty, Joseph Sangil Kwon
(713e) A Reduced Order Model for Optimizing Hydraulic Fracture Stimulation of Horizontal Wells 163
Cheng Cheng, Andrew P. Bunger
(713f) Prototype LIBS Sensor for Sub-Surface Water Quality Monitoring 164
Jinesh Jain, Dustin McIntyre, Daniel Hartzler
(730a) Catalytic Hydrogenation of Furfural over Rumop: Probing Bimetallic and Compositional Effects on Catalyst Performance 165
Yolanda Bonita, Timothy O'Connell, Jason C. Hicks
(730c) Etherification of 5-Hydroxymethylfurfural Using Zeolite Catalysts 166
Meredith Allen, Spencer Martell, Akbar Mahdavi Shakib, William M. Gramlich, Brian G. Frederick, Thomas J. Schwartz
(730d) Mechanistic Insights into the Hydrogenolysis of Levoglucosanol over Bifunctional Platinum Silica-Alumina Catalysts in Tetrahydrofuran Solvent 167
Siddarth H. Krishna, Rajeev Assary, Quinn A. Rashke, Zachary R. Schmidt, Larry Curtiss, James A. Dumesic, George W. Huber
(730e) Hydrothermal Stability of Chloromethyl Polystyrene Based Solid Acid Catalysts and Mechanism of Cellulose Hydrolysis 168
Maksim Tyufekchiev, Jordan Finzel, Pu Duan, Klaus Schmidt-Rohr, Sergio Granados-Focil, Marion Emmert, Michael T. Timko

(730f) Glycerol Transfer-Hydrogenation of Levulinic Acid Using Ru and Ir Carbene Organometallics Immobilized on Active Hydrotalcites 169
Jacob Heltzel, Kai Wang, Matthew Finn, Evan Sandefur, Adelina Voutchkova-Kostal
(730g) Paired Electrocatalytic Hydrogenation and Oxidation of 5-Hydroxymethylfurfural for Efficient Production of Biorenewable Monomers 170
Xiaotong Chadderdon, David Chadderdon, Wenzhen Li
(730h) A Full Furfural Utilization over Ni/SiO_2 Catalysts 171
Sheng-Chiang Yang, Shawn D. Lin
(182c) Two Stage Stochastic Programming Modeling Approach for Energy Storage System Operation Under Uncertainty 172
Jiah Yu, Jun-Hyung Ryu, In-Beum Lee
(733b) Theoretical and Experimental Investigation of the Microbial Degradation of Solitary Oil Microdroplets 173
George E. Kapellos, Nicolas Kalogerakis, Patrick S. Doyle
(733c) Modeling Faradaic Capacitive Deionization with Redox Active Porous Electrodes 174
Fan He, P. M. Biesheuvel, Martin Z. Bazant, T. Alan Hatton
(733e) Finding the Ideal Feedstock for Biodiesel Production Via Generalized Kinetic Model for Transesterification and Saponification 175
Pulkit Chhabra, Markus Kraft, Iftekhar A. Karimi, Anikesh Kumar
(733f) Accelerating the Generation of Coal Power Plant Property Models 176
Benjamin Sauk, Nick Sahinidis
(733g) Dynamic Optimisation of Water-Injection Wells Operation for Enhanced Oil Production from a Mature Oil and Gas Field 177
Emmanuel Epelle, Dimitrios I. Gerogiorgis
(733h) Voluntary Oil Well Ignition As a Blowout Response: Analysis of Factors Influencing Viability 178
A. Rashid Hasan, Joseph Sangil Kwon, Prashanth Siddhamshetty, M. Sam Mannan, Boyue Xu, Monir Ahammad
(740b) Integrated Forward Osmosis-Membrane Distillation Process for Sustainable Treatment of High Tds Produced Waters 179
S. Ranil Wickramasinghe, Peter Fyfe, Kamyar Sardari
(740c) Optimizing Sand Control in Wellhead with Advanced De-Sander Technology 180
Banchao Shu, Oluwatosin Oyelakin, Priyanka Shahi, Isaac Snyder, Mike Fredrick, Jeevan Dahal
Author Index

Some abstracts in this book refer to images that were not made available by the publisher

532378 Fabrication and Field Testing of Spiral-Wound Membrane Modules for CO_2 Capture from Flue Gas

Abstract:

A spiral-wound membrane module, including a spiral-wound membrane element, Plexiglas fiber-reinforced plastic tube, and membrane housing with face compression â€œOâ€ rings for effective seal of gases was developed. The spiral-wound membrane modules showed essentially no leakage and demonstrated an efficient gas separation. At a dry feed gas flow rate of less than 1000 sccm before humidification, a concentration polarization phenomenon was observed. Therefore, both dry feed and sweep gas flow rates each at 1000 sccm before humidification were used to evaluate the performances of the spiral-wound membrane modules at the typical flue gas temperature of 57°C. The feed and sweep gas pressures were at 1.5 and 1 psig, respectively. The spiral-wound membrane modules showed similar performance results with a CO_2 permeance of greater than 800 GPU and a CO_2/N_2 selectivity of more than 140 when tested with either a simulated flue gas or an actual flue gas. The results were similar to the results obtained from the flat-sheet membrane used for the module fabrication. The pressure drops of the modules were less than 1.5 psi/m and met the requirements for CO_2 capture from flue gas. The performances of the modules revealed some unexpected issues during the field test using the actual flue gas at the National Carbon Capture Center in Wilsonville, Alabama, including the feed spacer indentations on the membrane selective layer and the leakage at glue lines. Improvement of the spiral-wound membrane module fabrication was made by using a smoother feed spacer and a longer glue curing time. The spiral-wound membrane module was scaled up to a 5 inches diameter to accommodate a membrane area of around 14000 cm^2. A suitable spacer was selected to accommodate a spiral-wound membrane module test with vacuum at the permeate side. The field test results of the modules have shown their good potential for the post-combustion CO_2 capture from coal-fired power plants.

533205 Low Viscosity Water-Lean Diamine Solvents for Carbon Dioxide Capture

Abstract:

Carbon dioxide capture, storage (CCS), utilization and conversion of large anthropogenic CO_2 released from point sources such as coal fired power plants are possibly effective ways to minimize emissions. Toward this goal numerous CO_2 capture processes have been developed, the most mature of these technologies utilizes aqueous primary and secondary alkanolamines such as monoethanolamine (MEA) or diethanolamine (DEA) for chemical absorption of CO_2. Deployment of these aqueous amine solvents to market is limited by the high regeneration temperature due to the high heat capacity of water. Novel water-lean transformational solvent systems have been developed to address this limitation. We have combined multiple solvent chemistries and key findings from different solvent classes to develop a new class of water-lean diamine solvents with potential to be energy efficient CO_2 capture materials. These diamines bind CO_2 at standard temperature and pressure (STP) to form liquid carbamate salts with high gravimetric capacity and low viscosity without the need for co-solvents. Herein we will discuss the synthesis, characterization and testing of these novel amine solvents.

534889 High Swing Capacity MIL-101(Cr) Fiber Sorbents for Sub-Ambient CO_2 Capture Via RCPSA

Abstract:

A challenge facing CO_2 capture from flue gas via adsorption is the high cost of adsorbent materials, a phenomenon partially driven by low operating capacities leading to massive sorbent requirements. Increasing operating capacity can be accomplished by advances in sorbent materials, cycle design, and process design, giving the sorbent more capacity or making it operate more efficiently. This talk focuses on designing new structured sorbents using metal organic framework (MOF) materials to enable high capacity via a sub-ambient rapidly cycled pressure swing adsorption (RCPSA). We have found flue gas compression and cooling is counterintuitively energy efficient when extensive heat integration and energy recovery is utilized, and this concept enables high operating capacity for a variety of MOF sorbents. Current technoeconomic estimates reveal parasitic loads as low as 19% and total costs of CO_2 as low as 37$/tonne.

In this talk, after a brief review of the sub-ambient RCPSA process flow sheet, we will discuss methods of incorporating MIL-101(Cr) into fiber sorbent contactors via direct spinning. Sub-ambient CO_2 isotherms show MIL-101(Cr) may be capable of swing capacities as large as 10 mmol/g, making it an ideal material for sub-ambient RCPSA. Incorporating this material into a fiber sorbent contactors make possible order of magnitude lower pressure drops as well as enabling thermal management, which increases sorbent utilization. Phase change material incorporation into the fiber sorbent during spinning, and its effectiveness as thermal modulation will be presented. The talk will focus on the application of these fibers to capturing CO_2 from simulated flue gas, and will include analysis of breakthrough curves and preliminary cyclic results. These fiber sorbents containing high operating capacity sorbent materials help to support the case for applicability of the novel process design.

535795 Carbon Dioxide Capture By an Electrochemically-Mediated Amine Regeneration

Abstract:

Abstract

With the increasing evidence of global warming and its correlation to carbon dioxide (CO_2) emission, developing cost-effective, large-scale and efficient CO_2 capture technologies is critical. Thermal amine scrubbing is the most developed of these technologies, but is expensive to retrofit it into existing power plants. Recently, we demonstrated an electrochemically-mediated amine regeneration (EMAR) as a low-energy, efficient, and potentially scalable method of capturing CO_2 from flue gas [1, 2]. In an EMAR, the CO_2 rich stream is exposed to electrochemically generated copper ions (i.e., Cu = Cu^{2+} + 2eâ˘) to destabilize the reaction product of amines and CO_2 and release CO_2. Once the gas is flashed off, the lean loading stream is regenerated via the electrochemical plating of copper from the copper-amine complex (Cu^{2+} + 2eâ˘ = Cu). Therefore, electrochemical modulation of the concentration ratios of cupric ion and amine could afford a rather efficient, energy-saving and absorption-based CO_2 separation process.

The EMAR approach offers several advantages compared to other thermal driven amine-based processes of capturing CO_2. Unlike the thermal processes, EMAR can be operated at low temperatures, thereby minimizing thermal amine degradation. It also offers the possibility to desorb CO_2 at moderate pressures, leading to minimizing the downstream compression costs for CO_2 storage. The modular nature of EMAR, as an electrochemical cell, makes it easier to scale up and to operate with fluctuating loadings.

We studied the kinetics and thermodynamics of the EMAR process. In a rotating disk electrode and using the Tafel expression of the kinetics, we investigated several combinations of metals (the complexing agent), amines and background electrolytes. Among them, an EMAR system based on copper (i.e., Cu/Cu^{2+}), ethylenediamine, and sodium sulfate electrolyte was promising. In order to describe the thermodynamics of the system, twelve aqueous speciation reactions are solved simultaneously together with the extended Debye-Hückel activity coefficient model. These kinetics and thermodynamics data can give useful insight into designing an optimized system.

Since EMAR is a relatively new approach for capturing CO_2, there are still many opportunities to further improve the performance of the process. The initially developed EMAR used two identical copper plates as the electrodes. Recently, we studied the importance of the electrode by investigating different carbon-based electrodes (i.e., carbon paper, carbon cloth, carbon felt, and carbon foil). The results showed that the EMAR performance could be improved by using commonly-used and inexpensive carbon felts, mainly due to a lower charge transfer resistance, a higher surface area, and a faster nucleation rates.

The presentation will describe the EMAR process in detail and with updated performance results obtained in the lab-scale setup. The presentation will also highlight the challenges of the EMAR process as well as the various steps being undertaken in order to improve the efficiency of the process.

References

[1] M.C. Stern, F. Simeon, H. Herzog, T.A Hatton, Post-combustion carbon dioxide capture using electrochemically mediated amine regeneration, *Energy Environ. Sci.* 6, 2013, 2505-2517.

[2] M.C. Stern, T.A Hatton, Bench-scale demonstration of CO2 capture with electrochemically-mediated amine regeneration, *RSC Adv.* 4, 2014, 5906-5914.

540082 Layer-By-Layer Functional Thin Film Coatings for Enhanced Light Gas Separations

Abstract:

Current industrial hollow-fiber (HF) polymeric membranes generally employ high-flux, low-cost and low-selectivity polymeric membranes for these light gas (H_2, CO_2, CH_4) separations, while the majority of research to-date has focused upon identifying costly and exotic polymeric materials or composites of expensive polymers and ceramic powders, to achieve high selectivities at competitive separation rates. In contrast to this conventional approach, we aim to employ thin sub-micron conformal coatings, achieved via layer-by-layer (LbL) deposition, of functional polymers in order to improve the performance of existing low-cost and conventional HF membrane systems. This talk will present a review of our work employing layer-by-layer (LbL) assemblies for enhanced gas separations, specifically employing LbL films formed from (i) branched polyethylimine/polyacrylic acid (PEI/PAA) and (ii) polyethylene oxide/polymethylmethacrylate (PEO/PMAA), for selective hydrogen and carbon dioxide removal, respectively.

534659 Facilitated Transport Membranes for Hydrogen Purification from Coal-Derived Syngas

Abstract:

Facilitated Transport Membranes for Hydrogen Purification from Coal-derived Syngas

Yang Han and W.S. Winston Ho

William G. Lowrie Department of Chemical and Biomolecular Engineering, The Ohio State University, Columbus, OH

In this presentation, three amine-containing facilitated transport membranes were tailored for a single-stage membrane process to decarbonize coal-derived syngas. In these membranes, water-swellable crosslinked polyvinyl alcohol was used as polymer matrix, along with high MW polyvinylamine as fixed-site carrier. For Membrane 1, an amino acid salt of 2-(1-piperazinyl)ethylamine and sarcosine was used as the mobile carrier; sarcosine has a mild degree of steric hindrance. In Membrane 2, the 2-(1-piperazinyl)ethylamine salt of 2-aminoisobutyric acid (PZEA-AIBA) was used as the mobile carrier; AIBA is a sterically hindered amine. Membrane 3 utilized 1-(2-hydroxyethyl)piperazine as the mobile carrier, which contains 2 amino groups and a CO_2-phile hydroxyl group. In addition, poly(ethylene glycol) dimethyl ether was incorporated in Membrane 3 as a CO_2-phile moiety. Nanoporous graphene oxide was dispersed in all three membranes to avoid membrane compaction upon high feed pressure. The membranes were tested at 107 °C and 31.7 bar feed pressure with a CO_2 partial pressure of 12.5 to 0.5 bar, which corresponded to the feed CO_2 partial pressures before and after the bulk CO_2 removal, respectively. The separation performances of Membranes 1 and 2 depended strongly on the CO_2 partial pressure, yielding a CO_2 permeance of 72 – 435 GPU (1 GPU = 10^{-6} cm^3 (STP)$\cdot cm^{-2}\cdot s^{-1}\cdot cmHg^{-1}$) with a CO_2/H_2 selectivity of 93 – 558. The CO_2 permeance reduced significantly with increasing CO_2 partial pressure due to the carrier saturation phenomenon. On the other hand, the presence of amino groups and the abundant CO_2-phile moieties in Membrane 3 enhanced the CO_2 solubility at high CO_2 partial pressure, thereby a CO_2 permeance of 206 – 246 GPU and a decent CO_2/H_2 selectivity of 103 – 123 were obtained for a CO_2 partial pressure of 12.5 – 3.5 bar. The different degrees of dependence on the feed CO_2 partial pressure provide a design opportunity for a single-stage membrane process. The more permeable but less selective Membrane 3 can be implemented to the proximity of feed inlet, where the CO_2 partial pressure is high but the H_2 partial pressure is low. Upon CO_2 removal, the feed CO_2 partial pressure sharply reduces whereas the H_2 transmembrane driving force increases. In this case, Membrane 1 or 2 can be used to elevate the CO_2 permeance and CO_2/H_2 selectivity. Initial techno-economic analysis shows that the single-stage membrane process can achieve 90% CO_2 removal with >99% H_2 recovery. A 15.7% increase in cost of electricity is estimated, which is much cost-efficient then Selexol.

531546 Novel Membranes for CO_2 concentration and Carbon Capture

Abstract:

The capture of CO_2 from power plants and industrial exhaust gases could significantly reduce man-made CO_2 emissions. Widespread adoption of this technology would go a long way to solve the global climate change issue. Flue gas streams from power plants typically contain 4–20% CO_2 in mostly N_2 and are at atmospheric pressure. The objective of the separation process is to produce >99% pure liquid CO_2, at a pressure of 100–150 bars at a cost in the range of $40 – 60/ton of CO_2captured. This is a tough cost target for any process to meet. One of the best approaches seems to be to use a membrane system to perform a bulk separation that produces a CO_2 concentrate stream containing 40–80% CO_2. A membrane – assisted condensation process is then used to produce liquid CO_2. Because of prohibitive cost of compressing the flue gas feed stream to generate a driving force across the membrane, vacuum operation with a pressure of 0.2–0.3 bar on the permeate is the approach generally used. The pressure difference across the membrane is then no more than 1 bar, so very high permeance membranes must be used. Today's best membranes made from rubbery polar polymers have a CO_2permeance of about 2000 GPU and a mixture CO_2/N_2 selectivity of 20–30 at application conditions (30°C, water saturated). Increased selectivity is not particularly helpful because of pressure ratio limitations, but membranes with higher permeances would help reduce cost. For example, a membrane system for a typical 500 MW_e coal power plant is likely to have ~1 million m^2 of membranes, which would be a significant fraction of the plant cost. A doubling of membrane CO_2 permeance without selectivity loss would roughly cut the required membrane area in half, significantly reducing cost and increasing the competitiveness of the membrane process.

Compact Membrane Systems (CMS) has developed custom amorphous fluoropolymer (CAF) membranes for selective concentration of CO_2 from power plant flue gas and other industrial gases. Initial attempts to develop CMS CAF membranes for concentrating CO_2 are very encouraging (CO_2 permeance 4,500 GPU, CO_2/N_2 selectivity 20-30). The CAF membranes are also expected to have high chemical resistance and high anti-fouling capability due to their highly fluorinated nature. This presentation will discuss our promising membrane results plus downstream engineering evaluation.

532557 Process Development of Novel Sub-Ambient Membrane Processes for CO_2 Capture

Abstract:

CO_2/N_2 selectivity of membrane in the sub-ambient conditions is significantly increased, which allows considerable savings in energy consumption for membrane-based CO_2 capture process. In order to fully utilize benefits of high selectivity, we proposed novel membrane systems such that energy efficiency is maximized with full appreciation of capability in membrane for CO2 removal from coal-fired flue gas. The membrane process is configured to allow cost-effective use of membrane modules as well as energy-efficient generation of sub-ambient operating conditions through external refrigeration cycles. Enhancement of CO2 capture performance is made with the strategic use of gas sweeping to membrane permeate side. The proposed process is rigorously modeled and optimized in a holistic manner, subject to design specifications and engineering constraints, in which the selection of the most appropriate membrane network configurations and the determination of economic operating conditions is systematically made. Technical feasibility of new membrane process is assessed in collaboration with experimental group and membrane manufacturer, while economic evaluation in conceptual costing level is carried out to obtain CAPEX and OPEX, leading to the calculation of CO_2 capture costs. The optimized sub-ambient membrane process proposed allows about 15% reduction in CO_2 capture cost and 16% in parasitic load, compared to that of the conventional multi-stage membrane process operating at ambient conditions. Case study will be given to discuss considerable advances made in the new sub-ambient membrane process and to provide conceptual insights and design guidelines in the use of membrane for CO_2 capture process in practice.

Acknowledgement

This research was supported by the Korea Carbon Capture & Sequestration R&D Center (KCRC) through the National Research Foundation of Korea (NRF) funded by the Ministry of Science, and ICT (NRF-2014M1A8A1049338)

534912 U.S. DOE National Energy Technology Laboratory: Carbon Capture R&D Program's New Direction for Operational Versatility

Abstract:

U.S. DOE National Energy Technology Laboratory:

Carbon Capture R&D Program's

New Direction for Operational Versatility

José D. Figueroa[1], Lynn Brickett[1], and Kanwal Mahajan[2]

U.S. Department of Energy, National Energy Technology Laboratory

[1]626 Cochran Mill Road, Pittsburgh, PA 15236-0940, U.S.A

U.S. Department of Energy, National Energy Technology Laboratory

[2]3610 Collins Ferry Road, Morgantown, WV 26507-0880, U.S.A

ABSTRACT:

The Carbon Capture Program implemented by the U.S. Department of Energy's (DOE) Office of Fossil Energy and managed by the National Energy Technology Laboratory (NETL) supports the fundamental understanding and applied research development of a suite of transformational post-combustion and pre-combustion carbon capture technologies. These transformational capture systems and enabling technologies that promote favorable performance characteristics of capture systems must also address the technical and operational flexibility required to align with the changing energy profile of the United States. In 2016, CO_2 emissions from natural gas power generation surpassed coal emissions due to low-cost shale gas production. However, the cost of capture systems continued to represent approximately 75% of the total cost of carbon capture, utilization and storage (CCUS) systems that impact the economic opportunity for domestic deployment and export markets.

This presentation will discuss the breadth of the Department of Energy's Carbon Capture Program transformational carbon capture systems and enabling technologies, a new focus on optimal operational levels not relegated to 90% capture rates in order to achieve low capital and economic costs, and the latest status and lessons-learned from the largest and diversified capture project portfolio that also includes hybrid, traditional and non-traditional capture technologies.

538882 Highlights of Collaboration between the U.S.DOE Carbon Capture Simulation for Industry Impact Program and the Discovery of Carbon Capture Substances and Systems Initiative

Abstract:

The Department of Energy's (DOE) Fossil Energy (FE) Carbon Capture Program recently launched the Discovery of Carbon Capture Substances and Systems (DOCCSS) Initiative with the objective of commercializing transformational carbon capture technologies. This initiative integrates fundamental contributions from national laboratories with the practical perspective of industrial stakeholders to streamline the commercialization process of transformational carbon capture materials. To efficiently inform the design and operation of processes that fully leverage the unique benefits of the transformational CO_2 capture technology supported by the DOCCSS initiative, rigorous multi-scale modeling that links performance projections from the materials level through the device scale is critical.

The U.S. DOE's Carbon Capture Simulation for Industry Impact ($CCSI^2$) project is designed to utilize the open source computational tools and models developed under the Carbon Capture Simulation Initiative (CCSI) in partnership with industry to scale-up new and innovative carbon capture technology. Led by the National Energy Technology Laboratory (NETL), $CCSI^2$ operates in conjunction with and in support of the Department of Energy's (DOE) Fossil Energy (FE) Carbon Capture Program to focus on accelerating and reducing the risk of advancing promising technologies. The $CCSI^2$ technical team includes researchers from Lawrence Berkeley National Laboratory (LBNL), Lawrence Livermore National Laboratory (LLNL), Los Alamos National Laboratory (LANL) and Pacific Northwest National Laboratory (PNNL), West Virginia University, and the University of Texas at Austin.

$CCSI^2$ provides fundamental analysis, modeling, and optimization of carbon capture technologies by working closely with industry partners. In addition, $CCSI^2$ efficiently identifies data collection needs, characterizes carbon capture materials, designs and optimizes devices and processes, and propagates uncertainty in model predictions to provide a detailed perspective on the accuracy of results. $CCSI^2$ assists carbon capture technology developers by providing the following:

- More detailed understanding of capture materials through system performance under parametric uncertainty
- Designs for high performance and intensified unit operations
- Synthesis of processes optimized for novel materials
- Characterization of dynamic system behavior
- More informed design, operating & control decisions
- Optimized processes under uncertainty
- A framework for improved design of experiments at all levels of maturity to maximize learning to reduce risk.

The $CCSI^2$ framework for materials analysis through system optimization supports the DOCCSS Initiative recently launched by DOE/NETL. DOCCSS requires a multi-hierarchical characterization which embodies materials through systems level performance. The major contribution of $CCSI^2$ is to provide specific guidance on systems and device scale optimization to accommodate the unique characteristics of transformational carbon capture materials. Where necessary, through implementation of optimal design of experiments (DoE), $CCSI^2$ will also provide insight on additional data requirements that can maximize the information content in the data. Ultimately, a significant reduction is expected in development time and cost of materials and processes as compared with conventional approaches. Furthermore, an estimate of the uncertainty of all the key economic measures will be developed.

The collaboration between $CCSI^2$ and DOCCSS will enable more rapidly identification of optimal system/material pairing that can be commercialized with lower risk. We will present the most recent computational results developed from collaboration with the three recently awarded DOCCSS concepts:

1) Lawrence Berkeley National Laboratory (LBNL) Metal-Organic Framework (MOF) Sorbent Materials

2) Pacific Northwest National Laboratory (PNNL) CO_2 Binding Organic Liquids (CO_2BOL)

3) Lawrence Livermore National Laboratory (LLNL) Additively Manufactured Reactors

LBNL discovered a class of diamine-functionalized MOF materials that exhibit cooperative adsorption of CO_2, generating step-like CO_2 adsorption isotherms. While this breakthrough results in a much higher working capacity for CO_2, the rapid adsorption also results in a rapid liberation of adsorption heat. The unique characteristics of the LBNL MOF require a system that is capable of rapid heat removal if the material's benefits are to be fully realized; otherwise, the steep isotherm cannot be effectively utilized in practice, reducing overall performance. $CCSI^2$ is characterizing MOF performance and evaluating cost-optimal system designs.

PNNL has developed a low-viscosity, water lean solvent with a polarity swing regeneration process that is anticipated to have favorable CO_2 capture costs. $CCSI^2$ is developing a validated CO_2BOL solvent models to derive an optimal system using conventional equipment that accommodates the solvent/antisolvent pairing unique to the PNNL CO_2BOL technology.

LLNL is developing advanced reactor designs with customized geometries made realizable through advances in 3D printing, such as gyroid surfaces and hierarchical flow paths. This new capability creates the potential for high performance, multi-functional, intensified devices that can consolidate processes and result in lower capital cost. $CCSI^2$ is analyzing the potential to integrate the proposed PNNL CO_2BOL solvent and liquid anti-solvent system unconstrained by conventional equipment assumptions. The resulting optimal device design will be constructed by LLNL leveraging advanced manufacturing techniques. A computationally-efficient framework for exploration of novel geometry options will be developed.

Results of predictive LBNL MOF isotherm modeling of the "step-change" isotherm behavior, PNNL CO_2BOL systems level optimization, and development of the LLNL computational device design framework will be presented.

536624 Process Modeling and Experimental Studies of a Diamine-Appended Metal–Organic Framework for CO_2 Capture

Abstract:

Recent focus on climate change has driven research in the area of carbon capture, specifically in the area of post-combustion capture from power generation systems. Current commercial technologies for post-combustion CO_2 capture are mainly based on amine-based solvents, but they suffer from high energy penalty, degradation and low working capacity. Functionalized sorbents offer a potential alternative technology. However, viable sorbents should have lower regeneration energy, faster kinetics, higher CO_2 selectivity, and higher working capacity under the low partial pressure of CO_2 for 90% CO_2 capture from the post-combustion flue gas. A novel class of diamine-appended metal–organic frameworks (MOFs) offers a number of these features. These MOFs exhibit step-shaped isotherms under low partial pressures of CO_2 and lead to higher working capacities under similar temperature and pressure intervals when compared to traditional sorbents. The MOF studied in this work, Mg_2(dobpdc) ($dobpdc^{4-}$ = 4,4'-dioxidobiphenyl-3,3'-dicarboxylate), contains one-dimensional hexagonal channels with unsaturated Mg^{2+} sites and is functionalized by binding the diamine 2,2-dimethyl-1,3-diaminopropane (dmpn) to these sites which then chemisorb CO_2 in a cooperative mechanism to form carbamate and carbamic acid species. Along with high working capacities, this functionalized MOF also shows excellent long-term stability and maintains performance under humid conditions, both desirable attributes for flue gas CO_2 capture.[1]

For minimizing the penalty due to CO_2 capture, contactors for these functionalized MOFs should be optimally designed. The step-shaped isotherms exhibited by these MOFs cannot be adequately represented by traditional isotherms. It is suspected that significant physical adsorption and multiple chemical reactions occur simultaneously, and that multiple assumptions underlying well-studied traditional isotherms are violated in the actual mechanism of CO_2 uptake in this MOF class. A new isotherm model that captures the unique mechanism of CO_2 uptake is developed in this work. Data from thermogravimetric analysis (TGA) are used to develop the kinetic model. Using these newly developed models, a temperature swing adsorption system is optimally synthesized. This system targets a thermal management strategy that effectively manages the rapid heat transfer required to fully exploit the step isotherm behavior and realize the projected working capacity increases. Due to the cyclic nature of the process and due to spatial variation in the loading profile along the contactor, a multi-objective optimization problem is solved to minimize the amount of sorbent and energy required for post-combustion CO_2 capture.

References:

[1] – Milner, P.J., Siegelman, R.L., Forse, A.C., Gonzalez, M.I., Runcevski, T., Martell, J.D., Reimer, J.A., Long, J.R. A Diaminopropane-Appended Metal-Organic Framework Enabling Efficient CO_2 Capture from Coal Flue Gas via a Mixed Adsorption Mechanism. *Journal of the American Chemical Society.* 2017; 139 (38), 13541-13553

536649 Analysis of Different Solvent Performance in UKy-CAERâ€™s 0.7 MWe CO_2 Capture Pilot Plant

Abstract:

To mitigate costs associated with implementing post-combustion CO_2 capture (PCCC) technologies, and to make them commercially viable for application in electric power generation and utility plants, different research paths are continually being explored. To this end, UKy-CAER has developed and scaled up its PCCC technology at a 0.7 MWe scale integrated with a 750 MWe commercial coal-fired power plant at E.W. Brown Generating Station located in Harrodsburg, Kentucky in a project which was primarily funded by the U.S. Department of Energy, National Energy Technology Laboratory (DOE NETL).

In the UKy-CAER CO_2 technology, novel concepts are employed in a heat integrated process to maximize heat utilization to lower the parasitic energy of the capture process. These include, first, a two-stage stripping process for solvent regeneration for enhanced solvent capacity and a reduction in the associated energy for regeneration. Here, after the conventional primary steam-driven solvent regeneration, the lean solvent is sent to a secondary air-stripping column where air is used counter-currently to further lower the lean loading of the solvent providing a higher free-amine solvent to the top of the absorber. The higher CO_2 volume air from the top of the secondary stripper is recycled to the plant boiler as secondary combustion air, which in turn would increase the amount of CO_2 entering the absorber. The lower liquid phase CO_2 concentration from the leaner solvent and higher gas phase CO_2 concentration will increase the driving force for CO_2 diffusion through the liquid/gas reaction film with resultant higher mass transfer. Second, a heat integrated liquid desiccant-based cooling tower loop recovers waste energy from the CCS, improves power plant efficiency and reduces parasitic load. The benefits for economic savings are derived with the use of advanced solvents with the process.

In this work, the performance of two advanced solvents were evaluated to demonstrate the energy savings of the process. For the more energy efficient advanced solvent, a reduction of about 36% in regeneration energy was obtained relative to DOE reference case 10. The energy ranged from 900 â€“ 1500 Btu/Ib CO_2 from the parametric studies. The impacts of process parameters such as solvent circulation, lean loading, stripping temperatures, intercooling effects among others were examined. The heat integrated process is impacted by many process variables acting in tandem. Statistical analysis was used to determine the relative impacts of process variables and showed that intercooling effects in the absorber were more significant for the faster kinetics solvent from the observed temperature profile providing insights to the different behavior of the solvents and how the process could be operated to maximize the energy savings.

536630 *In-Situ* Gasification Chemical Looping Combustion vs. Chemical Looping with Oxygen Uncoupling: Exergy Comparison for Power Generation with CO_2 Capture

Abstract:

The chemical looping combustion (CLC) technology is one of the most promising CO_2 capture technologies for fossil fuel power generation applications, which allows effective separation of CO_2 from fuel combustion by avoiding contact between fuel and air. When applied to coal-based power plant, CLC can cost-effectively generate clean electricity with high net plant energy efficiency compared to amine-based post combustion CO_2 capture process. CLC refers to the use of a metal oxide or metal sulfate as an oxygen carrier to supply oxygen from air to the fuel through a reduction and oxidation reaction pathways of the oxygen carrier, respectively. Similar to oxy-combustion carbon capture, a CO_2 product stream is produced, undiluted with the N_2 inherently present in air. In contrast, the oxygen carrier redox reaction pathway replaces the need for an energy intensive air separation unit. This work focuses on two CLC process configurations of coal combustion: in-situ gasification chemical looping combustion (iG-CLC) and chemical looping with oxygen uncoupling (CLOU). CLOU process utilizes a copper-based oxygen carrier (OC) in a dual circulating fluidized bed reactor system consisting of 2 reactors: the reducer reactor, where the Cu-based OC releases O_2 for coal combustion, and the combustor reactor for OC regeneration by air. Both reactors and both flue gas streams can generate high quality heat for power production. In the iG-CLC process, an iron-based OC is circulated between a counter-current moving bed reducer reactor for full combustion of coal and a fluidized bed combustor reactor for OC regeneration by air. High quality heat can be extracted from reducer flue gas, combustor flue gas and combustor for power generation. The present work analyzes the iG-CLC and CLOU coal-based power plants to investigate the power generation potential of CLC technology as well as compare the pros and cons of both configurations. Specifically, this study developed the overall models of iG-CLC and CLOU coal-based power plants with ASPEN PLUS process simulation software. Previous bench-scale and sub-pilot experimental results performed at Ohio State were incorporated to iG-CLC model. CLOU model was constructed with reference to literature and report. Energy and material balance information generated by the models was utilized to perform detail exergy analysis on iG-CLC and CLOU configurations. The results showed exergy intensive point for process optimization. Parametric studies were also conducted on system pressure, reactor temperature, etc., to investigate important parameters of both CLC configurations.

519751 The Development of Machine Learning, Group Contribution and Molecular Modeling Approach to Screen Physical Solvents for Gas Separation

Abstract:

In this work, we focus on screening and designing solvents to physically and selectively absorb one gas from the other in a gas mixture. Specifically, a hybrid computational approach was developed to systematically screen solvents to separate CO_2 from a fuel gas mixture which contains CO_2, H_2 and H_2O. We have identified a solvent which is hydrophobic, non-foaming, less volatile and less viscous than the commercial Selexol solvent. Additionally, this solvent exhibits a large CO_2 loading and large CO_2/H_2 loading selectivity compared with Selexol. Due to its very high hydrophobicity, water presence in the gas stream will not affect CO_2 loading in this solvent. In contrast, water presence in the gas stream will adversely decrease CO_2 loading and CO_2/H_2 loading selectivity by 2-3 times in the hydrophilic Selexol. Along with other favorable properties for this solvent, such as the large overall CO_2 mass transfer coefficient obtained from simulations and physical model estimation and the reasonable price, this solvent is promising to be used in CO_2 pre-combustion capture applications and is being experimentally tested at NETL. Additionally, some solvents with much higher CO_2 loadings than the commercial methanol solvent used at low temperatures have also been identified. From our in-house database and molecular simulations, the largest CO_2 loading in any organic compound at 298 K was estimated to be 11 mol/MPa.L and the theoretical minimal CO_2 loading was approximately calculated to be 0.4 mol/MPa.L. A literature survey of experimental CO_2 loadings in ~100 compounds shows that most of those CO_2 loadings are between 0.4 - 11 mol/MPa.L; the largest experimental CO_2 loading obtained from the literature survey is about 3.3 mol/MPa.L, which is about 3.3 times smaller than the theoretical largest value of 11 mol/MPa.L. This result indicates that there is still much room for improvement to find better solvents with higher CO_2 loadings. Finally, some findings obtained from machine learning algorithms including artificial neural network will also be presented. It was found that the H_2 loading in a solvent correlates almost linearly with the solvent free volume fraction. For CO_2, a relationship between CO_2 loading and three other factors, that is, the solvent free volume fraction, effective CO_2-solvent interaction by including solvent molar volume, and functional availability was built. The best machine learning model gives a CO_2 solubility of about 0.4 mol/MPa.L different than the experimental data. Although more accurate models are being pursued, this preliminary model can be used for screening purposes to select solvents with large CO_2 loadings.

536637 Towards the Development of a Solvent Screening Tool for CO_2 Capture Using Molecular Thermodynamics

Abstract:

In the context of sustainable development and clean energy production, one of the most important alternatives to mitigate anthropogenic CO_2 emissions is to capture and separate CO_2 (CCS) from diluted sources, such as gases emitted from fossil fuel combustion and other industrial processes. In this regards, and given some of the limitations of current absorption processes, the development of new alkanolamines/amines is a topic which has attracted a great deal research interest. One of the major challenges when evaluating the techno-economic performance of new amines and blends is the lack of experimental data on the thermophysical properties, required for a reliable process design and simulation. Having a trustworthy modeling tool to fill the gap would be of high value. We have developed and implemented a robust theoretical framework for the description of key thermophysical properties of aqueous solutions of single and mixed alkanolamines/amines at relevant gas separation process conditions. The approach is based the combination of soft-SAFT [1], a molecular-based equation of state, with the Free-Volume Theory and the Density Gradient Theory for the integrated modeling of phase behavior, enthalpies, densities, viscosities and interfacial tensions. We present and discuss here the application of this procedure to investigate the absorption of CO_2 in five selected amines, differing in families and molecular structures: primary monoethanolamine (MEA), as a benchmark, secondary diethanolamine (DEA), tertiary methyldiethanolamine (MDEA), sterically hindered 2-amino-2-methyl-1-propanol (AMP) and cyclic piperazine (PZ). The study is performed in a systematic manner, starting from the development of the models for the pure amines [2], the description of their thermophysical properties, the properties of the aqueous mixtures [3] and the absorption of CO_2 in aqueous solutions of single and blended amines at conditions of relevance for post-combustion CO_2 capture applications. A scheme of implicit reactions is used to describe the formation of carbamate and/or bicarbonate products resulting from the chemical reactions between CO_2 and the examined amines. This procedure eliminates the need to specify the detailed equilibrium reactions and significantly reduces the number of parameters required to represent the absorption process. A maximum of two adjustable model parameters (one of which with a linear temperature dependency), optimized for a fixed amine concentration, suffice to represent the absorption of CO_2 in aqueous solutions of single amines over a broad range of temperatures (298 413 K) and CO_2 partial pressures (0.1 1000 kPa). The extrapolation capability of the model is tested by predicting the absorption of CO_2 in aqueous solutions of single amines for different amines concentrations, in reasonable good quantitative agreement with experimental data. The predictive capability of the model is evaluated for predicting the absorption of CO_2 in blends. The models are finally used to assess the CO_2 capture performance of selected amine systems in terms of solvent cyclic capacity and regeneration energy. These results demonstrate the feasibility of the developed approach as a reliable platform for the screening of amine solvents as function of key process parameters and as a valuable tool for process modeling and optimization, being also applicable to model other absorbents or processes.

This work has been funded by the Gas Research Center, The Petroleum Institute, Khalifa University of Science and Technology, through projects GRC17001 and GRC18003.

[1] F.J. Blas, L.F. Vega, Mol. Phys. 92, 135â€"150 (1997). F.J. Blas, L.F. Vega, Ind. Eng. Chem. Res. 37, 660â€"674 (1998).

[2] J.O. Lloret, L.F. Vega, F. Llovell, J. CO2 Util. 21, 521â€"33 (2018)

[3] L.M.C. Pereira, F. Llovell, L.F. Vega, Applied Energy, *in press* (2018)

533465 Systematic Design of Phase-Change Solvents for Post-Combustion CO_2 Capture Based on Advanced Thermodynamics and Holistic Sustainability Assessment

Abstract:

Solvent-based, post-combustion CO_2 capture in absorption/desorption processes is a mature technology, but its wider industrial adoption is severely challenged by high energy penalties in solvent regeneration and the environmental impacts associated with solvents and their derivatives. Phase-change solvents represent a class of materials promising to deliver significant energy reductions compared to conventional solvents (Wang and Li, 2015). They consist of miscible solute-solvent mixtures (e.g., amines in water) which undergo a liquid-liquid phase separation upon a change in the processing conditions, e.g. upon reaction of CO_2 with the solute or upon increasing the temperature after the reaction etc. This behaviour results in the formation of a CO_2-lean phase which may be separated through a non-thermal approach. The separation of the CO_2-lean phase reduces significantly the amount of CO_2-rich phase entering the desorber, while desorption may also take place at much lower temperatures than in the case of single-phase solvents. As a result, the required regeneration energy may drop below 2.5 GJ/ton CO_2, which is much lower than the 4 GJ/ton CO_2 required for conventional solvents such as Monoethanolamne (MEA).

Without exception, the few research efforts reported in recent years toward the identification of phase-change solvents have been based entirely on lab and pilot-scale experiments. This approach has resulted in the identification of very few phase-change solvent options that exhibit desirable regeneration energy reductions at the expense of very high experimental costs and without necessarily evaluating their sustainability performance. At the same time, it did not eliminate the existence of other options exhibiting similar or better behaviour. There are countless combinations of potential phase-change solvent and blend candidates and there is also a need for combined consideration of numerous thermodynamic, kinetic and sustainability properties as performance criteria prior to selecting phase-change solvents with optimum capture features.

To address these challenges, we approach for the first time the design of phase-change solvents through optimization-based Computer-Aided Molecular Design (CAMD). We have previously applied CAMD for the design of conventional CO_2 capture solvents with very promising results (Papadopoulos et al., 2016). However, the employed tools did not include indices and models for prediction of phase-change behaviour while data gaps in sustainability assessment could not be handled effectively. We now significantly extend this CAMD approach by introducing additional solvent design criteria that account for liquid-liquid phase separation, we incorporate an automated, holistic sustainability assessment framework which addresses data gaps within CAMD and we rigorously predict solvent-CO_2-water vapour-liquid-liquid equilibria using the group contribution SAFT-γ-Mie equation of state (EoS) (Papaioannou et al., 2014) for selected solvents. The proposed approach involves a screening stage where molecular structures are generated and evaluated during CAMD based on criteria and constraints such as CO_2 and water solubility in solvent, solvent vapour pressure, heat capacity, viscosity and basicity, to name but a few. Such properties capture the solvent effects on the process thermodynamic and reactivity performance, guiding CAMD toward useful options with the potential to exhibit liquid-liquid phase separation. In addition, we consider during CAMD numerous solvent properties associated with cradle-to-gate life-cycle assessment and safety, health and environmental hazard assessment. Such properties include cumulative energy demand for solvent production, global warming potential, mobility, fire or explosion, acute and chronic toxicity, water- and air-mediated effects, degradation in the environment and accumulation. To address the lack of models for the prediction of some of these models as well as the data gaps existing in several others we propose for the first time the use of a local lazy learning approach in the course of CAMD. This approach exploits structural similarities with a set of reference molecules to provide predictions for the desired properties. The employed CAMD method uses multi-objective optimization to identify few promising phase-change solvent options which are then evaluated through a rigorous equation of state. We employ the group-contribution SAFT-γ-Mie EoS (Papaioannou et al., 2014) to predict for the first time the vapour-liquid-liquid equilibrium behaviour of selected phase-change solvents. The employed EoS implicitly accounts for the phase and chemical equilibrium characteristics of solvent-water-CO_2 mixtures hence it predicts the mixture behaviour without the need to postulate reactions or reaction products. Being a group contribution approach, it does not require regression of pure component and mixture interaction parameters for the specific solvents. Solvents are represented as sets of functional groups for which parameters may be transferred from or to other molecules regardless of the application.

The proposed CAMD approach identifies phase-change solvents which exhibit favorable overall CO_2 capture performance, while we also consider solvent purchase price as an additional criterion for post-CAMD solvent analysis and selection. We investigate the impacts of considering sustainability indices during solvent design, as opposed to cases where sustainability is not considered. We find novel solvents, including cyclic and acyclic amines, for use as components in efficient phase-change mixtures and we also provide experimental verification for their suitability as CO_2 capture options. Known phase-change solvents such as N-Methylcyclo hexylamine (MCA) and N,N-Dimethyl cyclohexylamine (DMCA) (Zhang et al., 2012) are also designed during CAMD, hence verifying the validity of our approach.

Acknowledgements

This project has received funding from the European Union's Horizon 2020 research and innovation programme under the grant agreement 727503 - ROLINCAP – H2020-LCE-2016-2017/H2020-LCE-2016-RES-CCS-RIA.

Cited References

Papaioannou, V., Lafitte, T., Avendaño, C., Adjiman, C. S., Jackson, G., Müller, E. A., Galindo, A. 2014, Group contribution methodology based on the statistical associating fluid theory for heteronuclear molecules formed from Mie segments. Journal of Chemical Physics, 140, 054107–29

Papadopoulos, A.I., Badr, S., Chremos, A., Forte, E., Zarogiannis, T., Seferlis, P., Papadokonstantakis, S., Galindo, A., Jackson, G. and Adjiman, C.S., 2016. Computer-aided molecular design and selection of CO_2 capture solvents based on thermodynamics, reactivity and sustainability. Molecular Systems Design & Engineering, 1(3), 313-334.

Wang X., Li B., 2015, Chapter 1 - Phase-change solvents for CO_2 capture, In Novel Materials for Carbon Dioxide Mitigation Technology, edited by Fan Shi and Bryan Morreale, Elsevier, Amsterdam, 3-22.

Zhang J.F., Qiao Y., Agar D.W., 2012, Improvement of lipophilic-amine-based thermomorphic biphasic solvent for energy-efficient carbon capture, Energy Procedia, 4, 92-101.

540773 2018 Outlook for Energy: A View to 2040

Abstract:

The Outlook for Energy is ExxonMobil's global view of energy demand and supply through 2040. It also highlights the dual challenge of providing affordable energy to support prosperity while reducing environmental impacts – including the risks of climate change. We use the Outlook to help inform our long-term business strategies and investment plans.
The Outlook anticipates significant changes through 2040 across the world to boost living standards, reshape the use of energy, broaden access to abundant energy supplies, and accelerate decarbonization of the world's energy system to address the risks of climate change.

- Global energy needs will rise about 25 percent over the period to 2040
- Electricity demand will nearly double in non-OECD countries
- Worldwide electricity from solar and wind will increase about 400 percent
- Natural gas will expand its role, led by growth for electricity generation
- Growth in oil demand will be driven by commercial transportation and chemical needs
- Efficiency gains and growing use of less-carbon intensive energy sources will contribute to a nearly 45 percent decline in the carbon intensity of global GDP

Energy is fundamental to modern life, and as the world's population approaches 9 billion people in 2040, we are challenged to improve living standards everywhere. Understanding the factors that drive the world's energy needs – and likely choices to meet those needs – is the mission of the Outlook. By sharing the Outlook with the public, we hope to broaden that understanding. Energy matters to everyone, and we all play a role in shaping its future.

541543 Energy Decarbonisation Scenarios

Abstract:

With the energy sector being associated with 2/3 of the annual greenhouse gas emissions, decarbonizing the energy is a key step to reduce their level. The National Determined Contributions (NDC's) announced so far by the participating countries do not meet the objectives of the Paris-climate agreement, as far as reaching a peak in emissions in the near future or decreasing to a level that is compatible with the stated long-term targets.

Using different scenarios for energy demand and supply, options for energy sector decarbonisation are evaluated. Meeting the 2-degree target requires a major transformation in the power sector, efficiency improvements – specially in the buildings and in the industry, and a revolution in transport towards fuel and modal shifts. With a significant increase in renewable energy, system integration and optimization will also be key in reducing the cost of the energy transition. Meeting more ambitious climate goals ("well below 2 degrees") will push further the technological innovation and implementation requirements throughout the entire energy value chain. A scenario aiming at zero net-emissions by 2060 is presented, along with a comparison to the 2-degree case.

542625 Fundamental Research Needs to Advance Energy Technologies

Abstract:

The U. S. Department of Energy (DOE), Office of Science, Office of Basic Energy Sciences (BES) supports fundamental research in chemical and materials sciences to provide the foundations for new energy technologies and to support DOE missions in energy, environment, and national security. This presentation will discuss opportunities for fundamental research to impact DOE's energy mission "to catalyze the timely, material, and efficient transformation of the nation's energy system and secure U.S. leadership in energy technologies" with a focus on the way we generate, store and use energy nationally. I will provide an overview of BES strategic planning over the past decade that identified priority research directions for advancing energy applications, highlight key scientific advances in these areas, and discuss some future opportunities for modern science, particularly chemical sciences, to accelerate the transformation of the U. S. energy portfolio.

534275 An Overview of Recent Efforts Under the Department of Energy's Carbon Storage Program: Moving CCS Towards Commercialization

Abstract:

The Carbon Storage Program implemented by the U.S. Department of Energy's (DOE) Office of Fossil Energy and managed by the National Energy Technology Laboratory (NETL) is helping to develop technologies that safely and permanently store carbon dioxide (CO_2) without adversely impacting natural resources or hindering economic growth. Specifically, the Carbon Storage Program is developing and advancing carbon capture and storage (CCS) technologies both onshore and offshore that will significantly improve the effectiveness of the technology, reduce the cost of implementation, and be ready for widespread commercial deployment. Since 1997, DOE's Carbon Storage Program has significantly expanded the CCS knowledge base in selected technology areas through a diverse portfolio of applied research and field laboratory projects. The Storage program is comprised of three primary technology areas: (1) Storage Infrastructure, (2) Advanced Storage R&D, and (3) Risk and Integration Tools. These areas work together to address significant technical challenges in order to meet program goals that support the scale-up and widespread deployment of CCS.

Since 2003, DOE has been sponsoring research-oriented field projects through the Regional Carbon Sequestration Partnership (RCSP) Initiative. The Initiative consisted of three Phases: Phase I - Characterization (2003-2005); Phase II - Validation (2005-2011); and Phase III - Large Volume Injection Tests (2007 to 2022). As the RCSP Initiative nears completion, the Carbon Storage Program is moving forward with the Carbon Storage Assurance Facility Enterprise (CarbonSAFE) Initiative which is intended to research the feasibility of a commercial scale storage project (50 Mt+ storage over 25+ years). The CarbonSAFE Initiative began in FY16 and is also taking a phased approached starting with the Integrated CCS Pre-Feasibility phase followed by the Storage Complex Feasibility phase with the potential for additional phases in the future. Phase I, Pre-Feasibility, is nearing completion and Phase II, Feasibility, is underway. This presentation will include the latest updates from the CarbonSAFE Initiative.

536327 Chemical Impacts of CO_2 Intrusion into Heterogeneous Caprock

Abstract:

Carbon capture, utilization, and storage (CCUS) is considered one of the most practical options for reducing CO_2 emissions from stationary sources. Particularly, CO_2-enhanced oil recovery (CO_2-EOR) with CO_2 injected into oil reservoirs is a well-established and relatively mature technology that offers potential economic benefits of increasing oil production and reducing costs of CO_2 storage. Sealing formation (caprock) integrity is considered as a key risk factor, because of the potential for leaked CO_2 or brine to migrate into shallow groundwater formations. The purpose of this work is to evaluate the long-term caprock sealing capacity and effects of caprock hydrological heterogeneities on CO_2 migration. The Farnsworth Unit (FWU), an oilfield in the northern Anadarko Basin in Texas undergoing active CO_2 enhanced oil recovery (EOR) is selected as a case study. A two-dimensional reactive transport model of the caprock (Morrow Shale and Thirteen Fingers) of the FWU was developed based on the site geological model. Specific objectives of this study include: (1) to understand how CO_2 is likely to influence the caprock integrity; and (2) to analyze the effects of caprock heterogeneity that control CO_2 migration and CO_2-water-rock interactions.

Heterogeneities in permeability and porosity could significantly affect containment of intruded CO_2 within the caprock. Key results of analysis include forecasted spatial and temporal CO_2 distribution in caprock near injection wells, because the pressure and CO_2 saturation tend to be the highest over the reservoir. The risk of CO_2 leakage through the caprock is evaluated, and mineral alteration of the caprock and self-sealing effects are assessed.

534634 Implementing and Validating Active Reservoir and Brine Management Strategies for the Enhancement of the Geologic Storage of Carbon Dioxide: An Update

Abstract:

The Energy & Environmental Research Center (EERC) is conducting a multiyear field demonstration to evaluate active reservoir management (ARM) strategies that have the potential to improve the commercial viability of the geologic storage of carbon dioxide (CO_2) by managing reservoir pressure, reducing stress on sealing formations, controlling the footprint of the injected fluid within a formation, and improving overall injection performance. This effort, the brine extraction and storage test (BEST), is being funded by the U.S. Department of Energy (DOE) National Energy Technology Laboratory, in partnership with Nuverra Environmental Solutions (Nuverra), Schlumberger, and Computer Modelling Group, and is being implemented at a field site located in North Dakota.

The BEST consists of two complementary components: 1) an ARM test and 2) a brine treatment test bed. Simultaneous injection and extraction of brine combined with a monitoring program will be used to evaluate ARM as a means of enhancing the geologic storage of CO_2 in a deep saline formation (DSF). The treatment of the formation fluids extracted as part of the ARM test will take place in a pilot-scale brine treatment test bed facility to investigate the capability of emerging water treatment technologies to treat high-salinity brine for beneficial use. Together, these two components will serve as a proxy for future saline CO_2 storage projects employing ARM to manage formation pressure plumes, subsurface CO_2 migration, and the extracted subsurface brine.

Phase 1 of the BEST project, which focused on final site selection and development of a field implementation plan, was completed in 2016 and yielded the following results:

- ARM testing will occur in the Inyan Kara Formation, a regionally extensive DSF previously identified as a prime target for commercial-scale CO_2 storage because of its superb geologic properties. This fact, combined with the presence of commercial brine injection operations at the Johnsons Corner site (Watford City, North Dakota) that emulate commercial-scale CO_2 storage (volumetrically equivalent to the injection of >250,000 tons/yr CO_2), makes the results of ARM in the Inyan Kara directly applicable to widespread adoption by carbon capture and storage (CCS) projects in DSFs.

- Recent and ongoing salt water disposal (SWD) operations at the Johnsons Corner site have created an ideal reservoir environment for conducting a BEST. Long-term operations at the site via two Class II underground injection control (UIC) injection wells have created a differential plume that can be modified through ARM, and the established injection history allows for confident simulated predictions.

- A four-well design, using the two existing SWD wells, provides operational flexibility and monitoring capability to test ARM scenarios through a range of injection and extraction rates. Injection can be independently controlled into both SWD wells, and the extraction rate, via a newly installed extraction well, can be varied. A new disposal well will also be installed into a lower geologic horizon for disposal of extracted water as a parallel to ARM implementation at CCS sites.

- While the native brine is expected to have salinities below 180,000 mg/L TDS, the produced water brines being delivered to the test site have a salinity typically greater than 300,000 mg/L TDS. Thus integration of the extracted brine treatment test bed facility with the operations of the Johnsons Corner SWD operations provides the ability to generate tailored brine concentrations and achieve the DOE target level of 180,000 mg/L TDS for the extracted brine treatment technology demonstrations.

Phase II of the project is now under way at the Johnson Corner site and consists of the execution of the field implementation plan, which comprises a site infrastructure design and implementation plan, a site operation plan, a monitoring plan, and a water treatment technology evaluation, design, and implementation plan. Currently, the necessary permits have been secured from the state of North Dakota and McKenzie County and the installation of an extraction well completed in the Inyan Kara Formation; an extracted brine disposal well completed in the Broom Creek Formation; and the brine test bed facility, brine-handling equipment (e.g., storage tanks, pipeline, etc.), and support infrastructure (e.g., additional power lines, access roads, etc.) are in progress. The brine test bed facility will provide a controlled environment to evaluate brine treatment technology performance for a range of beneficial use applications on a range of tailored brine compositions. Treatment rates ranging from 1 to 25 gallons per minute will be maintained over extended operating periods of 30 to 60+ days, with a desired minimum period sufficient to encompass a minimum of two maintenance cycles. Monitoring of all chemical and energy requirements will be performed along with the generation of all secondary residuals, sufficient to close a system energy and material balance.

The selection of the brine treatment technologies for Phase II testing is under way using a screening and selection process developed during Phase 1. This process consists of a scoring system that weights the areas of treatment cost, readiness level, safety considerations, and waste generation on a scale of 1 to 10. The selection process will give priority to those technologies associated with DOE-funded projects awarded for the development of innovative high-TDS (total dissolved solids) brine pretreatment and desalination technologies (DOE Funding Opportunity Announcements 0001095 and 0001238) that successfully satisfy technology screening criteria.

This presentation will provide an overview of the BEST Phase 1 project results and an update of the status of the Phase II ARM and brine treatment technology tests.

532007 Evolution of Transport and Mechanical Properties of Mt Simon Sandstones Due to Interaction with Brine/CO_2

Abstract:

When sandstone rocks are exposed to CO_2-saturated brine, their transport and mechanical properties can change due to brine/CO_2-induced chemical reactions. The present study investigates the change in the flow-through characteristics, porosity and the mechanical behavior of Mt. Simon Sandstone samples caused by exposure to brine/CO_2. The cores, extracted from the Mt. Simon formation, were first characterized for their mechanical and transport properties, and then aged in CO_2-saturated brine at pressures of 2500 psi and temperature of 50â°C for one to two weeks. Following that, tests were carried out to measure the change in transport and mechanical properties of the samples. Our experiments show that the porosity of the Mt. Simon samples slightly increases after the exposure to CO_2/brine. On the other hand, the permeability increases quite substantially (depending on the confining environment) upon exposure to CO_2/brine. Industrial CT and Micro CT scanning were used to monitor the evolution of the pore structure during exposure. Measurements of the flow-through pore size distribution (PSD) are indicative of changes occurring consistent with the observed increases in permeability. Nitrogen adsorption tests (BET), before and after incubation, show a significant loss of pore volume in the mesoporous range that is indicative of clay dissolution. Weakening of the materials was observed based on the mechanical properties studied, a result that is consistent with the observed dissolution of clays that play a central role in the cementation of the quartz grains. The compositions of the brines used in the experiments were analyzed, and increases in concentration of most cations were found after incubation with the Mt. Simon cores, also consistent with mineral/clay dissolution, which is in line with the porosity, transport, and mechanical property measurements and electron microscopy analysis of the same samples.

532261 Optimization of CO2-Enhanced Oil Recovery with CO2 Storage in a Mature Oil Field

Abstract:

This paper presents an optimization methodology for CO2 enhanced oil recovery in mature oil reservoirs. A field-scale compositional reservoir flow model was developed for assessing the performance history of an active CO2 flood and for optimizing both oil production and CO2 storage in the Farnsworth Unit (FWU), Ochiltree County, Texas. A geological framework model constructed from geophysical, geological, and engineering data acquired from the FWU was the basis for all reservoir simulations and the optimization method. An equation of state was calibrated with laboratory fluid analyses and subsequently used to predict the thermodynamic minimum miscible pressure (MMP). Initial history calibrations of primary, secondary and tertiary recovery were conducted as the basis for the study.

After a good match was achieved, an optimization approach consisting of a proxy or surrogate model was constructed with a polynomial response surface method (PRSM). A sensitivity analysis was first conducted to ascertain which of these control variables to retain in the surrogate model. The PRSM utilized an objective function that maximized both oil recovery and CO_2 storage. Experimental design was used to link uncertain parameters to the objective function. Control variables considered in this study included: water alternating gas cycle and ratio, production rates and bottom-hole pressure of injectors and producers. Other key parameters considered in the modeling process were CO_2 purchase, CO_2 recycle and adding infill wells and/or patterns as well as compressor capacity. The PRSM proxy model was â€˜trainedâ€™ or calibrated with a series of training simulations. This involved an iterative process until the surrogate model reached a specific validation criterion. A genetic algorithm with a mixed-integer capability optimization approach was employed to determine the optimum developmental strategy to maximize both oil recovery and CO_2 storage.

The proxy model reduced the computational cost significantly. The validation criteria of the reduced order model ensured accuracy in the dynamic modeling results. The prediction outcome suggested robustness and reliability of the genetic algorithm for optimizing both oil recovery and CO2 storage. The reservoir modeling approach used in this study illustrates an improved approach to optimizing oil production and CO2 storage within partially depleted oil reservoirs such as FWU.

This study may serve as a benchmark for potential CO2â€“EOR projects in the Anadarko basin and/or geologically similar basins throughout the world.

525249 How Much Oil and CO_2? at What Cost? Bounding the Quantities of Produced Oil and Stored CO_2 from CO_2 EOR in Conventional Oil Fields in the United States

Abstract:

Carbon dioxide (CO_2) enhanced oil recovery (EOR) is a technology that uses supercritical CO_2 as a solvent to extract additional oil from a conventional oil field after oil production from primary production and secondary water flooding are no longer economical. If captured CO_2 is used in CO_2 EOR, then the CO_2 that remains in the subsurface after CO_2 EOR operations conclude is stored CO_2. There are many conventional oil fields in the United States where CO_2 EOR could be applied. The National Energy Technology Laboratory (NETL), which is part of the Office of Fossil Energy (FE) within the United States (US) Department of Energy (USDOE), has developed two models for rapidly evaluating potential oil production and CO_2 storage resulting from application of CO_2 EOR. The FE/NETL CO_2 Prophet Model is a streamline/stream tube reservoir simulator that models oil production and CO_2 storage resulting from application of CO_2 EOR. The output from the FE/NETL CO_2 Prophet Model is used as input to the FE/NETL Onshore CO_2 EOR Cost Model which is a cash flow model that estimates the profitability of the oil, water and CO_2 flows by computing all of the associated revenues, capital costs, operating costs, royalties, taxes (including federal income taxes), financing costs and, ultimately, the net present value of earnings after taxes. This model also calculates the break-even oil price that provides investors with their minimum desired return on investment for the oil field. The Energy Information Agency (EIA), which is part of USDOE, maintains a dataset of geologic and oil properties for many of the conventional oil fields in the lower 48 states in the US. After screening the oil fields in this dataset for suitability for CO_2 EOR, we exercise the models to find that approximately 40 billion barrels of oil are technically recoverable from these oil fields. At an oil price of $75/barrel, 16 billion barrels of oil are economically recoverable with roughly 9 billion tonnes of CO_2 stored. However, there is significant uncertainty associated with many of the variables used in the analysis. In this paper, ten oil fields from different regions of the US are selected and subjected to a sensitivity analysis where geologic properties and economic variables are systematically varied to illustrate how changes in these variables affect the break-even price of oil. Based on this analysis, selected variables are adjusted and applied to the entire oil field dataset to illustrate how uncertainty in these variables affects the calculated volume of oil produced and mass of CO_2 stored using CO_2 EOR at different market prices for oil.

530155 Ultra-Deep Desulfurization of Low-Sulfur Gasoline By Selective Adsorption of Trace Mercaptans over Supported Metal Oxides

Abstract:

The removal of trace amount of refractory sulfides in gasoline has increased the cost of fuel processing. In low-sulfur gasoline, trace amount of mercaptans is commonly present yielded from the reversible reaction of dissolved hydrogen sulfide from hydrotreating with olefin in gasoline. Developing effective final clean-up process for the removal of trace amount of mercaptans from low-sulfur gasoline at low energy consumption can be a complementary strategy for hydrotreating. In this work, we proposed a novel continuous selective adsorption-reactive desorption process for ultra-deep desulfurization of mercaptan from low-sulfur gasoline under ambient conditions. A series of supported metal oxides was screened for selective adsorption of heptanethiol from model fuel. With the optimized metal loading of 14%, the supported adsorbent showed super-high adsorption capacity of 31mg-S/g-sorbent at the initial heptanethiol concentration of 500 ppm-S. The adsorbent can be fully regenerated readily involving reactive desorption strategy under gentle conditions, with almost no loss on desulfurization capacity after three consecutive adsorption-regeneration cycles. Real gasoline desulfurization tests in fixed-bed further confirmed the effectiveness of the supported adsorbent to clean up the trace amount of mercaptans from gasoline under ambient conditions. The proposed facile continuous selective adsorption-reactive desorption process can be energy- and cost- effective for ultra-deep desulfurization of mercaptan from low-sulfur gasoline under ambient conditions.

532676 Catalytic Adsorptive Desulfurization (CADS) of Diesel Using Industrial-Grade MCM-41

Abstract:

ABSTRACT

The catalytic adsorptive desulfurization (CADS) has been developed as a promising approach for ultra-deep desulfurization of diesel with low energy consumption (Ren, X.; Miao, G.; Xiao, Z.; Ye, F.; Li, Z.; Wang, H.H.; Xiao, J. Catalytic Adsorptive Desulfurization Using TiO_2/SBA-15 under Mild Conditions. Fuel, *2016*, 174, 118-125; Ren, X.; Liu, Z.; Dong, L.; Miao, G.; Liao, N.; Li, Z.; Xiao, J. Dynamic Catalytic Adsorptive Desulfurization of Real Diesel over Ultra-stable and Low-cost Silica Gel Supported TiO2. AIChE J., 2018, doi:10.1002/aic.16055). In this work, we report our recent interesting experimental results on the effective CADS of diesel using economic industrial-grade MCM-41, despite of the ineffectiveness of reagent-grade MCM-41 at all. The CADS of industrial-grade MCM-41 shows high desulfurization capacity and selectivity, similar as those of our previously reported TiO_2/SBA-15. The adsorbent was characterized by FTIR, XRD and N_2 adsorption, and further clues on CADS mechanism using industrial-grade MCM-41was under further investigation in our lab. Effect of O/S, CADS temperature and selectivity were optimized in batch reactor. Spent adsorbents was regenerated by solvent wash followed with air oxidative treatment. Stable desulfurization performance of the adsorbent in 9 consecutive adsorption-regeneration cycles suggested the active sites on the adsorbents can be fully regenerated stably. The discovery in this work provides a facile route using the low-cost industrial-grade MCM-41 directly for CADS with superior desulfurization capacity and remarkable stability, which paves an effective and economic way for ultra-deep desulfurization of real diesel with low energy consumption.

536315 Evolution of CO_2 Storage Capacity Associated with Geochemical Reactions in Subsurface

Abstract:

Carbon dioxide (CO_2) can be stored in pore spaces in the subsurface via many trapping mechanisms, including hydrodynamic trapping, residual trapping, solubility trapping, and mineral trapping. Total storage capacity is closely related to storage formation properties, particularly porosity and permeability. In this study, we investigate the evolution of storage capacity during and after CO_2 injection. In particular, we evaluate the variation of porosity and permeability of storage formation caused by geochemical reactions between fluids and rocks. The Morrow B sandstone at the Farnsworth Unit (FWU) enhanced oil recovery with CO_2 (CO_2-EOR) field in northern Texas was selected as a case study. We quantify the impacts of reactive transport on CO_2 storage at FWU by comparing simulation results with or without geochemical reactions. We also evaluate the importance of including the reactive transport component in different CO_2 storage scenarios, i.e., CO_2-EOR and deep saline (assuming zero oil saturation at FWU). We hypothesize that the impact of reactive transport on altering porosity and permeability becomes insignificant when the oil phase presents in the storage scenario.

536850 Quantitative Analysis of the Influence of Capillary Pressure on Geologic Carbon Storage Forecasts. Case Study: CO_2-EOR in the Anadarko Basin, Texas

Abstract:

Numerical models are critical for forecasting subsurface multi-phase flow associated with geologic carbon storage. Uncertainty of model results stems from many factors, including and especially uncertainty in multi-phase flow parameters. Specifically, relative permeability and capillary pressure relationships depend on both the rock properties and fluid properties, and the latter may be highly nonlinear as fluid temperature and pressure conditions change. Forecasts of trapping mechanisms, phase behavior, and plume movement are impacted by the choice of relative permeability and capillary pressure functions and how those functions are calibrated and constrained. In particular, one of the most neglected aspects of such model simulations is meaningful capillary pressure processes, most likely due to lack of measured data. A primary goal of this study is quantify the difference in forecasts or results for models that utilize capillary pressure functions calibrated with measured data from results of models without such. Additionally, the relative permeability models developed here were derived from measured capillary pressure data. Those data were used to constrain saturation end points in the relative permeability curves and dictate how that relative permeability was distributed spatially.

The case study area for the research presented is the Farnsworth Unit (FWU), an active CO_2 â€" enhanced oil recovery field in the Anadarko Basin of northern Texas. Correlation among porosity and permeability data within the FWU suggest a series of definitive hydrostratigraphic units, zones of the reservoir that exhibit similar flow properties. In a separate study, capillary pressure was measured by mercury intrusion testing of 46 core samples from two different wells in the FWU. Thirteen of those 46 samples were correlated to specific hydrostratigraphic units within the principal reservoir, the Morrow â€˜Bâ€™ Sandstone. Five of the 13 samples were chosen for this study to represent each of the five hydrostratigraphic units identified in the reservoir. The relative permeability curve assigned to each of these flow units included saturation end-points determined from the capillary pressure data for that particular flow unit, in effect a heterogeneous parameter assignment. Results of models parameterized with this approach were compared to models that utilize a more conventional approach of assigning a single relative permeability and capillary pressure relationship to a reservoir, typically based on lithology type or geologic formation. Also compared were results of models parameterized with other synthetic relative permeability relationships paired to measured capillary pressure relationships and simulations run without capillary pressure.

The main conclusions drawn from this analysis include (1) heterogeneity in relative permeability plays a major role in simulated forecasts of CO_2 migration, trapping mechanisms and storage capacity, as well as oil and water production, and (2) capillary pressure, and in particular the magnitude of capillarity, also plays a major role in these processes; conversely, if the magnitude of capillarity is small relative to in situ fluid pressures, it imparts insignificant effects on these processes.

538628 Measurement and Calibration of Self-Sealing Rate of Fractures in Geological CO_2 Storage: Case Study of a Natural Analog

Abstract:

Injection of anthropogenic CO_2 into a subsurface reservoir for sequestration will impact the geochemistry, porosity and permeability of the reservoir significantly. If a fault or fracture penetrates the reservoir, CO_2-laden brine may migrate into that fault, eventually sealing it via precipitation or opening it up via dissolution. Having a complete understanding of such effects is crucial for successful deployment of CO_2 sequestration.

We studied the Little Grand Wash Fault-zone (LGWF) in central Utah, where outcrops evidence historical leakage of naturally occurring CO_2 causing self-sealing in fractures along the fault-zone. Moreover, the site also has active leakage of CO_2 and CO_2-enriched groundwater in the form of a cold-water geyser, leading to active precipitation of travertine ($CaCO_3$) on the surface. Thus, this site serves as an ideal natural analog for leakage along fault-zones from engineered CO_2 systems.

Our hypotheses were that (1) the precipitation reactions would be rapid enough to be measurable within one cycle of the geyser, and that (2) fractures in a fault-zone similar to the LGWF would self-seal with $CaCO_3$ at the measured precipitation rates. To test these hypotheses, we conducted a 24hr field reactor experiment at the geyser. The mean precipitation rate at the geyser at 24hr was estimated to be 2.9×10^{-3} mol/m^2/s. We then set up a reactive transport model for vertically upward flow through an idealized single fracture with conditions similar to the LGWF. Significant reduction of porosity and permeability was observed in the fracture regions where the precipitation rates matched to those measured in the fields. A key implication from this study is that CO_2-enriched waters that are highly saturated with $CaCO_3$ can potentially seal a near-surface fracture within tens of years. However, the permanency of such sealing may be dependent on the subsequent flow conditions.

543546 New Data and Models to Avoid Cryogenic Solids Formation in LNG Production

Abstract:

The formation and deposition of solids during the cryogenic processing of natural gas is a perennial risk for operators. Current tools for predicting heavy hydrocarbon solid formation temperatures suffer from various limitations including restricted composition ranges and/or an inability to represent the available literature data. New measurements of the trace BTEX, CO_2 and H_2O solubilities in multi-component LNG mixtures are essential if more robust predictive models are to be developed. Furthermore, the process gas chromatographs used to protect cryogenic heat exchangers from solids freeze-out do not have the resolution needed to distinguish high- and lower-risk compounds (e.g. benzene vs hexane), and only provide an indirect indication of risk: the composition measured must still be converted to a freeze-out temperature using a thermodynamic model.

We present here outcomes from our ongoing program to help avoid cryogenic solids formation in LNG production. A new software tool called ThermoFAST has been developed and endorsed by GPA Midstream to replace the Kohn-Luks Solids Solubility Program. High-pressure visual cells operating at cryogenic temperatures have been constructed and used to produce new solid-liquid equilibrium data to further validate and extend ThermoFAST. These tools have been used to explore phenomena of retrograde solidification in LNG systems, which may lead to novel methods of mitigating inadvertent solids formation in operating plants. Finally, our efforts to develop a more robust sensor capable of directly assessing heavy hydrocarbon freeze-out risk will be described.

539231 An Innovative Technology for the Production of Value Added Chemicals Using Fischer Tropsch Synthesis

Abstract:

INFRA Technology uses a novel Cobalt catalyst supported on Al_2O_3-based composite substrate for the production of paraffinic hydrocarbon from pipeline quality natural gas. Our commercial catalyst designated as â€œS1â€ produces mainly a mixture of gasoline and middle distillate fractions in 60/40 proportion. Within middle distillate, up-to 45% is kerosene fraction good for use in the aviation industry. The full spectrum of this technology is based on a GTL (gas to liquid) chemical process which entails a purification section for the removal of contaminant sulfur from pipeline quality gas. Further, heavier hydrocarbons are removed using a pre-reforming process. The main reforming process entails an environmental friendly tri-reforming process, which in addition to conventional wet reforming, also converts greenhouse carbon dioxide to a valuable syn-gas. Above 85% conversion is achieved in a single reformer, an endothermic reactor with 16 tubes. Commercially tested Nickel catalyst used for this purpose. The reforming process achieves hydrogen to carbon monoxide product at a ratio of 2.8:1, with carbon dioxide as a by-product. This by-product carbon dioxide from the reformer is recycled back to reformer using the conventional amine contactor-regenerator system. Less than 1% carbon dioxide slips through amine and goes to synthesis section as an inert. Syngas ratio correction is achieved via a membrane unit that utilizes the high solubility and diffusivity of hydrogen within the membrane fibre. Ratio corrected syngas (H_2 to CO at 2.1:1) is fed to the first stage of FT reactor. Approximately 60% conversion is achieved in the first stage of exothermic reactors. After inter-stage cooling and product removal, the reactants are left over syngas is fed to the second stage to achieve an overall conversion of 80%. FT reactors are fixed bed, shell and tube type of reactor that utilizes water-steam phase equilibrium for precise control of reactor temperature. This unit also eliminates the flaring by using the product gas as fuel to the process. It is also self-sufficient in meeting the process water demand from the product water. Thus it reduces the load on water table and improves ground sustainability. The application of INFRAâ€™s technology are enormous, including monetization of stranded gas reserves, conversion of refinery flare gas to value added chemicals, improving shale gas economics and providing a useful add-on in conjunction with any bio-gasification technology for BTL (biomass to liquid) commercialization. INFRA technology is also undergoing rapid innovation in the technology design and catalyst development. The aim is to further simplify the reactor design and improve catalyst yield, selectivity and stability. Further ENCON (energy conservation) projects are underway to improve the overall energy efficiency of the unit.

532432 Strongly Coupled Co@CoO$_x$ Nanoparticles and Layered Perovskite As a Highly Stable and Efficient Cathode for Solid Oxide Electrolysis Cells

Abstract:

Due to the increased concern of energy supply and environmental issues, the development of renewable energy technologies with low emissions is urgent.[1] Hydrogen with the highest specific energy density is considered as the best alternative to fossil fuel towards sustainability of energy.[2] Solid oxide electrolysis cells (SOECs) have received great attention owing to efficient hydrogen production from electricity at elevated temperature from both thermodynamic and kinetic perspectives, compared to low temperature electrolyzer.[3] However, conventional Ni based electrode suffers from poor redox-stability and coarsening during steam electrolysis.[4] Thus, the development of novel fuel electrode materials with high activity and stability is essential for practical application.

In this work, a stable and catalytically active layered perovskite $LaBaMn_{1.6}Fe_{0.2}Co_{0.2}O_{5+\text{Î´}}$ (LBMFC) with socketed Co@CoO$_x$ nanoparticles has been fabricated via *in situ* annealing nominal $La_{0.5}Ba_{0.5}Mn_{0.8}Co_{0.1}Fe_{0.1}O_{3-\text{Î´}}$ in 5% H_2 at 850 °C. A structure transformation (mixed cubic and hexagonal to layered tetragonal phase) and exsolution of core-shell Co@CoO$_x$ has been verified with thermogravimetric analysis, rietveld refinement of x-ray diffraction result and transmission electron microscopy with the energy-dispersive x-ray spectroscopy elemental mapping. The loss of lattice oxygen in this phase transition process leads to an increase of oxygen vacancy in LBMFC, consisting with the high concentration of adsorbed oxygen species, which is beneficial to electrical conduction and catalytic activity. A 300-Î¼m electrolyte supported electrolysis cell with Co@CoO$_x$ decorated LBMFC cathode operated in 10% H_2O-90% H_2 exhibits outstanding current densities of 1135.43, 744.19, 487.79 and 281.43 mA cm^{-2} under an applied voltage of 1.3 V at 800, 750, 700, 650 °C, respectively. Relevant hydrogen production rate of 511 ml cm^{-2} h^{-1} calculated from the Faradayâ€™s law has been achieved under an electrolysis voltage of 1.3 V, and the cell polarization resistance is as low as 0.125 Î© cm^2 under open circuit voltage at 800 °C. Moreover, the SOEC with LBMFC as cathode shows remarkable stability over 200 hours at a constant electrolysis current of -200 mA cm^{-2}, and the *in situ* exsolved Co nanoparticles are still embedded on the surface of the LBMFC framework without agglomeration observed by scanning electron microscopy, which suggests the strong connection between the exsolved nanoparticles and matrix. The increased oxygen vacancies together with the *in situ* exsolved high catalytic active Co@CoO$_x$ on the perovskite backbone generate sufficient active sites and consequently improves the performance and stability of LBMFC for the steam electrolysis.

531171 Bimetallic Pd-Cu Catalysts for CO_2 Hydrogenation to Methanol

Abstract:

Bimetallic Pd-Cu Catalysts for CO_2 Hydrogenation to Methanol

Xiao Jiang[1,2], *Xiaowa Nie*[2,3], *Xinwen Guo*[2,3], *Krista Walton*[1], *Chunshan Song*[2,3*]

[1] *School of Chemical and Biomolecular Engineering, Georgia Institute of Technology, 311 Ferst Dr., GA 30332, USA*

[2] *Clean Fuels & Catalysis Program, PSU-DUT Joint Center for Energy Research, EMS Energy Institute, The Pennsylvania State University, 209 Academic Projects Building, PA 16802, USA*

[3] *PSU-DUT Joint Center for Energy Research, School of Chemical Engineering,*

Dalian University of Technology, Dalian 116024, China

Email: csong@psu.edu

Utilization of CO_2 as a carbon source for synthesizing chemical feedstocks and transportation fuels has recently attracted great attention worldwide. The present work involves Pd-Cu bimetallic catalysts for CO_2 hydrogenation to CH_3OH and seeks to clarify if the combination of Pd and Cu could synergetically promote CH_3OH formation from CO_2/H_2 and to develop a fundamental understanding on the composition-structure-alloy-catalytic property relationship.

A strong synergetic effect on promoting methanol formation was observed over Pd-Cu bimetallic catalysts supported on amorphous silica when the Pd/(Pd+Cu) atomic ratios lied within 0.25-0.34 mol mol^{-1}. The optimal CH_3OH STY and selectivity were obtained on Pd(0.34)-Cu/SiO_2 (e.g., 0.31 Î¼mol g^{-1} s^{-1} and 34 mol%, respectively), which was almost 3 times as much as the simple summation of those over the corresponding monometallic catalysts. To examine the origin of the synergetic promotion, the reduced catalysts were characterized by TPR, XRD, TEM, STEM/EDS, and H_2-/CO_2-TPD. The results demonstrated that the metallic phases varied with the Pd/(Pd+Cu) atomic ratios, and uniform alloy phases $PdCu_3$ and PdCu, were found at Pd/(Pd+Cu) atomic ratio of 0.34 with an average crystallite size around 5 nm. TPD results indicated that the combination of Pd and Cu significantly promoted the adsorption towards weakly-bonded H_2 and CO_2 on the catalyst surface, the amount of which appear to correlate with the improved CH_3OH formation rate. The correlation of alloys with the activity performance indicates that the coexistence of nano-sized and uniform Pd-Cu alloy particles, $PdCu_3$ and PdCu, played a crucial role in the observed bimetallic CH_3OH promotion.

The catalytic performance was also studied at lower metal loading but maintaining the optimal bimetallic composition. Not only did the metal efficiency-based CH_3OH yield well retained when decreasing the metal loading from 18.7 to 2.4 wt%, but it also exhibited almost 2-fold more CH_3OH than the commercial Cu-ZnO-Al_2O_3 catalyst. Albeit such advantage, the Pd-Cu catalysts with lower metal loading yielded more CO, which is attributed to the Cu-rich alloy surface. Evidently, the product distribution strongly depends on the alloy composition.

The alloy composition-dependent behavior was investigated by both computational and experimental work, wherein PdCu alloy composition is superior for both CO_2 and H_2 activation and subsequent hydrogenation to CH_3OH in comparison to $PdCu_3$ alloy composition.

These findings would be of critical importance for developing new catalysts and catalytic CO_2 hydrogenation process technology and for fundamental understanding of chemistries involving in CO_2 hydrogenation to CH_3OH

536111 Energy Analysis of Non-Aqueous Solvents (NASs) for CO_2 Capture Process

Abstract:

Energy analysis of non-aqueous solvents (NASs) for CO_2 Capture Process

Aravind V. Rayer, Paul Mobley, Vijay Gupta, Jak Tanthana, Mustapha Soukri, Marty Lail and S. James Zhou

***RTI International**, 3040 E. Cornwallis Rd., Research Triangle Park, NC 27709-2194, USA*

Abstract

To develop a cost-effective absorption-based CO_2 capture process, it is necessary to evaluate the energy consumption of the scrubbing process which is the large contribution to the cost. This task needs the detailed solvent properties such as reaction enthalpy, Vapor-Liquid Equilibrium (VLE) and Liquid heat capacity. Investigation of both aqueous and non-aqueous solvents were done in this work based on a short-cut method using the in-house measured properties. The non-aqueous solvents reduced the heat of vaporization due to the low water content and high boiling points at the regeneration temperatures. The sensible heat contribution to the energy was reduced by the lower specific heat capacities of NASs and the higher working capacities for CO_2 capture.

In this work, the comparison of aqueous MEA with NASs showed that heat of absorption of NASs are lower than aqueous MEA as shown in Figure 1 and NASs are having higher working capacities as shown in Figure 2. The heat of absorption of NASs decreased with increase in temperature, whereas for aqueous systems heat of absorption increased with increase in temperature as shown in Figure 1. The effect of increasing reactive amine concentrations, water and alcohols on the heat of absorptions, heat capacities and VLEs were investigated. Increasing the amount of hydrophobicity of amine for NASs were also investigated. Specific reboiler duty (SRD) calculated using short-cut method showed that 1.94 MJ/kg_{CO2} could be achievable with an optimized NAS as shown in Figure 3.

Figure SEQ Figure * ARABIC 1. Comparison of heat of absorption of aqueous and non-aqueous system

Figure SEQ Figure * ARABIC 2. Comparison of Vapor-Liquid Equilibrium (VLE) of aqueous and non-aqueous

solvent

Figure SEQ Figure * ARABIC 3. Comparison of Specific Reboiler Duties (SRDs) of aqueous and non-aqueous system

538362 Microwave Assisted Lignin Depolymerization Using Deep Eutectic Solvents

Abstract:

Lignin, the second most abundant polymer, is greatly underutilized. Recent research has shown that lignin has a great potential for synthesizing wide variety of value added chemicals and renewable materials. However, one of the major hurdles in utilization of lignin is its complex polymeric structure. Lignin is a complex hetero-biopolymer composed of three different monomers (monolignols; H, G and S) connected through as many as eight different bonding motifs. The most common bonding motif in soft (~50%) and hardwood (~65%) lignin is the β-O-4 linkage. Lignin depolymerization is a crucial step in conversion of lignin to chemicals. Thermochemical and biochemical pathways for lignin depolymerization have numerous disadvantages such as high energy requirement to achieve high temperatures and limited product selectivity. Ionic liquids (IL's) have also been tested for lignin fractionation. While IL's have high efficiency, they are costly and sometimes toxic. A new line of solvents known as Deep Eutectic Solvents (DES) are a promising alternative for ionic liquids that can be used in lignin fractionation. DES's are cheaper than most ILs due to low precursor cost, simple synthesis and recyclability. DES is a mixture of two or more chemicals acting as either hydrogen-bond donors (HBD) or hydrogen-bond acceptors (HBA). The interactions between HBD and HBA of the DES provide a dual acid–base catalysis mechanism which will facilitate controlled cleavage of labile ether linkages among phenylpropane units and thus lead to lignin depolymerization. This interaction can be exploited to fractionate lignin and generate a low molecular weight lignin products. Due to rapid and volumetric heating, microwave based chemistry has significant advantages over conventional heating techniques. In this report we investigate the application of microwave heating for lignin depolymerization using various deep eutectic solvents. In this study we investigate three different DES's namely; oxalic acid: choline chloride (1:1), lactic acid: choline chloride (2:1) and Formic acid: Choline chloride (2:1). The DES's are synthesized in a 2450MHz microwave batch reactor. 10% by weight lignin sample is mixed with 30 g DES and the mixture is heated at different temperatures (110, 120, 130, 150, 170 °C) and times (1, 5, 10, 15, 20 min). The optimum time and temperature conditions will be determined and these conditions will be tested on conventional heating for comparison. The dissolved lignin is precipitated and characterized using GPC to compare the average molecular weight of the fractionated lignin products with original sample. GCMS and NMR analysis will also be conducted to understand the bond cleavage and product composition.

536443 Enhanced Gasification Reactor Designs for Maximizing Gas-Particle Interaction

Abstract:

Coal gasification is an endothermic chemical conversion process that converts coal and steam into syngas, a flammable mixture of hydrogen and carbon monoxide. However, the conversion efficiency of current gasification reactor technologies can be limited due to lack of gas-particle interactions within the reactor volume. To improve upon the current reactor technologies and maximize conversion within a given reactor volume, there is need to develop a reactor geometry that improves upon the fluidized bed and transports reactors without the issues of slugging, bypassing, and particle agglomeration that limit conversion.

In this work we discussed an effort to develop new fluidized bed gasification reactor designs that target maximization of reactorâ€™s mixing and resulting syngas yield. More than 15 unique reactor designs were outlined and investigated using reactive-CFD simulations of the gasification process. The reactive-CFD simulations undertaken using MFiX - multi-phase simulation package enabled assessment of the gas-particle mixing behavior and syngas (CO and H2) yield in each of the reactor designs. A reactor design optimization task was undertaken to maximize the mixing and yield. Subsequently a lab-scale prototype of novel reactor design was fabricated using additive manufacturing techniques. Cold flow experimental tests were undertaken to validate the MFix simulation flow pattern and pressure drop. The results show an improvement of 3 % in production of syngas in a small scale reactor design of 140 cm height and 7.5 cm inside diameter. Such designs could be easily adopted as radically engineered modular systems, which have the potential to provide novel, cost-effective gasifiers.

Acknowledgements: The work presented here was funded under a Department of Energy Phase I SBIR contract DE-SC0013822. The authors would like to thank Dr. Arun Bose of National Energy Technology Laboratory. The authors would also like to thank Dr. Sarma Pisupati from Pennsylvania State University who participated as a consultant and supported on baselining the gasification conditions. The simulations under this project were undertaken using high-performance computing infrastructure of NSF Extreme Science and Engineering Discovery Environment (XSEDE) under allocation TG-SBR150001 of Dr. Srujan Rokkam.

Keywords: Coal gasification, Syngas, fluidized bed, MFix, reactive-CFD, mixing, additive manufacturing

521361 Modeling the Decomposition Kinetics of the Gas Hydrates in Porous Medium

Abstract:

The hydrates present in the onshore and offshore reservoirs are the potential source of the natural gas for the future energy demand. Destabilizing these hydrates by the thermal stimulation, depressurization and inhibitor injection lead to recovering the entrapped natural gas. To extract the natural gas at a commercial scale, one needs to model the complex decomposition kinetics of the gas hydrates. In this view, we propose a rigorous kinetic model that can predict the methane gas recovery during the hydrate dissociation in the presence of a porous medium. This framework consists of a novel driving force (i.e., the difference in chemical potentials of water in the hydrate and liquid phase) and the reasonable reaction interface area (i.e., the effective surface area of the porous medium). In estimating the chemical potential of water in hydrate phase, we have used a rigorous approach[1] that incorporates the effect of movement of water molecules of the hydrate lattice, the guest-water interaction beyond first shell and the guest-guest interactions. Optimal values of the process parameters are determined by applying the generalized reduced gradient nonlinear optimization technique. The predictions from the proposed simulator are compared with the experimental[2] methane recovery in three systems that differ by the amount of silica sand and the operating temperature of the dissociation process. For the concerned situations, the proposed model has predicted the methane gas recovery with the absolute average relative deviation (AARD) in the range of 4.83-6.50%. This formulation can be significant to interpret the production potential of the natural gas hydrates present in the unconsolidated sand reservoirs.

References

[1] Ravipati S. and Punnathanam S. N. Improving the rigor and consistency of the thermodynamic theory for clathrate hydrates through incorporation of movement of water molecules of hydrate lattice. The Journal of Physical Chemistry C 119(22) (2015) 12365-12377.

[2] Linga P., Haligva C., Nam S. C., Ripmeester, J. A., Englezos P. Recovery of methane from hydrate formed in a variable volume bed of silica sand particles. Energy & Fuels 23(11) (2009) 5508-5516.

523755 Modeling Phase Equilibrium with Wong-Sandler Mixing Rule for Ternary $CO_2/H_2/C_2H_6$ Hydrates

Abstract:

Gas hydrate deposits in marine and permafrost region are believed to meet the worldwide energy demand for several decades. Besides energy production, the gas hydrate technology has plentiful industrial applications, e.g., carbon dioxide capture, desalination, gas storage, separation, transportation and so on. Hydrates are formed by encapsulating the small-sized molecules (guests) into the three-dimensional network of host (water) cages. The structures of the hydrates are specific to the interactions between guest and host molecules. Moreover, in case of mixed hydrates, a structural change is possible at different vapor phase composition (Babu *et al.*, 2013). The presence of polar and asymmetric components may impart nonideality in the fluid phases, for which, the linear mixing rules are inappropriate at high and low density limits. In this regard, we introduce the Gibbs free energy based Wong-Sandler (WS) mixing rule (Wong and Sandler, 1992) in association with the Patel-Teja (PT) equation of state model. The Gibbs free energy is calculated using the non-random two-liquid (NRTL) activity coefficient model. The three phase (vapor-liquid-hydrate) equilibrium is estimated by the van der Waals-Platteeuw model, in which the Kihara cell potential method is chosen to calculate the cage occupancy. The model is validated with a ternary mixture hydrate of carbon dioxide, hydrogen and propane, which forms both sI and sII hydrate at different compositions, 0.8/0.188/0.012 and 0.381/0.594/0.025 mole fractions, respectively (Babu *et al.*, 2013). The binary interaction coefficients are estimated by minimization of the average absolute relative deviation (AARD) in predicted and experimental pressure. The particle swarm optimization technique (*particleswarm* tool in MATLAB) is used for the same. Analysing the model performance, it is observed that the proposed PT-WS shows a better prediction (4.7% AARD) than the conventional van der Waals mixing rule based model (14% AARD). An accurate prediction of phase equilibrium of multicomponent hydrates using the proposed model can be useful in the hydrate based applications.

References

Babu, P., Yang, T., Veluswamy, H. P., Kumar, R., Linga, P. (2013). Hydrate phase equilibrium of ternary gas mixtures containing carbon dioxide, hydrogen and propane. The *Journal of Chemical Thermodynamics*, 61, 58-63.

Wong, D. S. H., Sandler, S. I. (1992). A theoretically correct mixing rule for cubic equations of state. *AIChE Journal*, 38(5), 671-680.

536056 The U.S. Department of Energy's R&D Program for Carbon Use and Reuse

Abstract:

The U.S. Department of Energy Office of Fossil Energyâ€™s National Energy Technology Laboratory has an active carbon capture and storage (CCS) program. A focus area within the carbon storage program is Carbon Use and Reuse. The objective of this area is to identify and develop a suite of technologies to beneficially use CO_2 to produce useful products excluding enhanced oil recovery technologies. A key challenge is to identify candidate chemistries and economically feasible approaches that utilize meaningful amounts of CO_2 as feedstock for in-demand products and yet have smaller greenhouse gas footprints than traditional pathways. Life cycle analysis (LCA) is being used to determine the environmental impact of CO_2 utilization projects. An open source LCA software, openLCA, is being used in the development of these LCAs. Current projects in the portfolio include cements/concretes, algae, and novel concepts.

539323 Enabling a Solid-State Carbon Dioxide Distribution Network

Abstract:

Efforts like the Carbon Capture, Utilization, and Storage R&D program implemented by the U.S. Department of Energy (DOE) Office of Fossil Energy are focused on developing technologies to capture, separate, and store CO_2 so that greenhouse gas emissions are reduced without adversely decreasing the U.S. energy supply and subsequently hindering economic growth.

In this presentation, a joint effort between RRTC Inc. and Rutgers University towards demonstrating the feasibility of a solid-state CO_2 capture, storage, distribution, and utilization (ss-CCSDU) paradigm is described. This "carbon network" is enabled by two novel process technologies. The first technology is known as low-temperature solidification (LTS), which can be used to capture CO_2 to form solid carbonates in an energy-efficient manner. The second technology that is key to the proposed paradigm is a hydrothermal release process that is normally used to synthesize submicron multicomponent oxide powders but can also be used as a low-energy route for separating CO_2 from solid carbonates in a controlled manner.

Together, the two technologies provide cyclability for capture and release of CO_2, allowing the transport of CO_2 in solid form from the site of its emission for utilization elsewhere. Calculations show that logistical management of CO_2 in this manner can be less costly and allows much more throughput than traditional storage forms like compressed gas and liquid CO_2. Such benefits are of great interest to those who utilize CO_2 in industrial processes.

534859 Plasma and Fluidic Oscillation Assisted Electrolysis of CO_2

Abstract:

The reduction in carbon emissions required for a sustainable future, and the resultant necessary decarbonisation of energy generation, inevitably lead to an increased focus on renewable energy sources. Natural intermittency of renewable electricity sources such as wind and solar, mean that technologies such as energy storage must play an increasingly fundamental role by smoothing the natural fluctuations in electricity production. Reversible Solid Oxide Cells (SOCs) are widely seen as a leading technology for future clean power generation, chemicals production, and energy storage. Renewable electricity can be utilised directly via electrolysis to reduce CO_2 and/or H_2O which can then be further reacted to produce a myriad of hydrocarbon related products. In times of low or no renewable electricity generation, the SOC can flexibly switch to run as a fuel cell to produce electricity.

Whilst single SOCs are easy to operate on a small scale in the laboratory, larger systems have found it difficult to compete with alternative energy technologies on cost, performance and durability. In particular, it is necessary to develop methods for lifetime extension of SOCs, minimisation of losses such as concentration polarisation, and faster chemical activation of CO_2, using energy inputs close to the thermodynamic minimum.

We are developing a novel, hybrid, plasma and fluidic oscillation assisted electrolysis system, in which the plasma is used to radically improve the kinetics and energy efficiency of CO_2 dissociation. The system is designed to reduce concentration polarisation by disrupting the gas boundary layer using fluidic oscillation and by utilizing ionic wind formed in plasmas (the gas flow generated by movement of ions in the plasma).

Non-thermal plasma catalysis has shown great potential for CO_2 reduction in its own right due to the promotion of strongly endothermic reactions with low activation energy, so that little or no excess energy is required from the plasma for activation and thermodynamic efficiencies are high. The challenges are to dynamically control the reaction and to achieve high conversion.

Fluidic oscillation can disrupt boundary layer formation and therefore minimise, or remove completely, the effects of concentration polarisation on cell performance. Fluidic oscillation has never before been coupled to an SOC.

Reactor design and preliminary results will be presented, showing the effect of fluidic oscillation and plasma on CO_2 electrolysis.

537821 Treatment and Extraction of Metals from Electronic Wastes Using a Novel Solvent Containing Supercritical CO_2

Abstract:

The management of wastes, particularly electronic waste (e-waste), has become a significant challenge faced by humanity. The number of discarded computers, phones, and other appliances, often defined as Waste Electrical and Electronic Equipment (WEEE), doubled between 2009 and 2014 to 41.8 million tons per year globally. Currently, e-waste disposal factories tend to recycle the non-toxic, valuable components of e-waste (metals) and release the toxic components into the surroundings via pyrometallurgy (burning). Chemical treatment, known as hydrometallurgy, is an alternative method that involves large amounts of hazardous, corrosive acids. As more and more e-waste is landfilled and disposed, the availability of easily mineable elements, particularly metals, decreases, disturbing the overall material cycle. Therefore, new, environmentally-friendly approaches to recycle and recover valuable components from e-waste are urgently needed. This study focuses on the use of supercritical CO_2 ($scCO_2$) and piranha acid as a novel solvent system for the treatment of e-waste, specifically in the extraction of metals, such as copper, iron and aluminum. Printed circuit board (PCB) was used as the e-waste of study and is generally composed of 40% metals, 30% plastics, and 30% refractory material. Through the synthesis of metal and polymer model composite particles, which reduced the complexity of PCB, the effects of $scCO_2$ in the solvent system were investigated. The results suggested that the role of $scCO_2$ is not the direct extraction of metal from e-waste particles but rather swelling of the polymer (organics) component within the e-waste particles. The swelling of polymers induced permanent morphological changes, and thus, seems to have enhanced the transport phenomena and ultimately improved the kinetics of the metal leaching into the solvent. A transport model was developed for the metal extraction from the complex e-waste particles consisting of layers with distinct organic and metallic components and a kinetics study was performed to gain a fundamental understanding of the multiphase interactions between the $scCO_2$, acid, and metal/polymer particle matrix. The $scCO_2$ solvent system was found to be a promising treatment scheme for metal extraction from e-waste with improved environmental sustainability.

535687 A CO_2 Utilization Approach Towards the Synthesis of Terephthalic Acid

Abstract:

Terephthalic acid (TPA) is the monomer used in the large polyethylene terephthalate (PET) plastics market (>30 MT/yr demand) and is currently made industrially by paraxylene oxidation (high-temperature, oxygen-based AMOCO process). Alternatively, TPA can be made from other precursors by using CO_2. The environmental concerns of CO_2 release into the atmosphere from traditional energy sources have led to significant interest in capturing and utilizing CO_2 as a chemical feed stock. RTI has developed a homogeneously catalyzed reaction to make TPA at low temperature (70Â°C) and high yield (>90%) with a simplified purification protocol. We are currently exploring new feed stocks, such as renewable biomass, to produce TPA at temperatures and pressures that are competitive with the AMOCO oxidation process, offering a CO_2â€“utilizing alternative to high-demand PET plastics in which CO_2 is sequestered long term in commonly known everyday products.

526826 Catalytic CO_2 Hydrogenation to C_2^+ Hydrocarbons

Abstract:

The catalytic CO_2 hydrogenation to higher hydrocarbons has attracted increasing attention due to the environmental and industrial requirement in mitigating CO_2 emissions and the need for reducing dependence on petroleum-based chemicals and fuels [1]. The past efforts in the field of CO_2 hydrogenation focused on modifying Fe-based catalysts that are used for Fischer-Tropsch (FT) synthesis with the addition of K or Mn promoters [2,3]. Our recent studies have demonstrated a new promising direction of research on Fe-based bimetallic catalysts such as Fe-Co tailored for direct CO_2 hydrogenation [4,5] to C_2-C_7 hydrocarbons as well as the C_2-C_4 light olefins.

In our most recent investigation, Fe-Cu bimetallic catalysts supported on γ-Al_2O_3 with a wide range of Cu/(Cu + Fe) atomic ratios were prepared. Hydrocarbon synthesis activity was tested in a fixed-bed flow reactor at 573 K under 1.1 MPa using 24% CO_2/72% H_2/4% Ar as feed gas after H_2 reduction at 673 K. Results show that upon combining Cu and Fe together, the CO_2 conversion increases with the increasing Cu content with the maximum at Cu/(Cu + Fe) atomic ratio = 0.17, but CO_2 conversion decreases with a further increase in Cu content beyond this ratio. The STY of C_2−C_7 reveals an identical trend as that for CO_2 conversion and maximizes at Cu/(Cu + Fe) atomic ratio of 0.17; however, CH_4 is significantly suppressed by the increase of Cu loading, while CO is promoted in the meantime. Considering Cu is a reverse water gas shift catalyst (RWGS) to generate CO but cannot produce hydrocarbons, Fe−Cu bimetallic catalysts show strong synergetic promotion in the higher hydrocarbon synthesis. The Anderson−Schulz −Flory (ASF) distribution was also examined for the chain growth. At the optimal composition (i.e., Cu/(Cu + Fe) = 0.17), the alpha value also maximizes, indicating the promoting effect of Fe−Cu combination on carbon-carbon chain growth. These results reveal that the Fe-Cu bimetallic catalyst with a small amount of Cu demonstrated a strong bimetallic promotion on CO_2 conversion and C_2^+ hydrocarbons selectivity [6]. The addition of Cu to Fe successfully suppresses the undesired CH_4 formation while enhancing CO and C_2^+ hydrocarbon formation.

The effect of K addition on catalytic performance of Fe−Cu bimetallic catalysts was also studied. The results show that K addition to Fe−Cu gives higher CO_2 conversion and C_2−C_7 STY at all the Cu/(Cu + Fe) atomic ratios than the corresponding unpromoted Fe−Cu catalysts, while at the same time, CH_4 is suppressed after adding K. At lower loading level of K, although paraffins still dominate, light olefins are observed evidently. C_2−C_4 light olefins are selectively produced and become dominant among light hydrocarbons when the K/Fe atomic ratio was increased to 0.5 and 1.0. Such K loading-dependent behavior becomes more distinct when plotting the STY of light olefins (C_2−C_4) and corresponding olefin/paraffin ratio (O/P) as a function of K/Fe atomic ratio. Clearly, the addition of K into Fe−Cu bimetallic catalysts significantly enhances the formation of light olefins as the STY increases with the increase of K/Fe atomic ratio. In addition, the corresponding alpha values of these K promoted Fe−Cu catalysts display a similar K loading dependent behavior as both C_2−C_4 olefin STY and O/P ratio. Hence, the addition of K into Fe-Cu dramatically enhances the production of olefin-rich C_2-C_4 hydrocarbons over Fe-Cu bimetallic catalysts. The Fe-Cu/K catalysts exhibit superior selectivity towards C_2^+ hydrocarbons synthesis than Fe-Co/K catalysts under the same reaction conditions.

XRD results of the reduced catalysts suggest the Fe-Cu alloy formation in Fe−Cu(0.17)/Al_2O_3 catalyst, which is in accordance with the DFT calculations by Nie et al. [7,8], that adding Cu to Fe can create alloy-like species, and thus leads to a completely different pathway without going through CO as the intermediate product, which has been considered the dominant pathway for Fe monometallic catalyst. We also find the shifts of H_2 peaks in Fe-Cu bimetallic catalysts compared with Fe and Cu monometallic ones from H_2-TPR tests, which demonstrate the surface species transformation after adding Cu to Fe. Furthermore, the reduction degrees of Fe-Cu bimetallic catalysts differ from those monometallic ones physically combined in proportion, which, again reveals the new adsorption site formation on the Fe-Cu bimetallic catalysts, compared with Fe monometallic one. The CO_2 adsorption property is also proved to be affected by adding Cu into Fe, by testing CO_2-TPD. Larger peaks appear around 500K in Fe-Cu bimetallic catalyst, and there exist stable adsorbed CO_2 from 600 to 700K only in Fe-Cu bimetallic catalysts, which corresponds to both CO_2 conversion and C_2^+ hydrocarbon promotions. More characterization tests are ongoing to explore the relationship between physico-chemical properties of the Fe-Cu bimetallic catalysts with the synergetic effects in CO_2 hydrogenation to hydrocarbons.

References

[1] Song, C., Global challenges and strategies for control, conversion and utilization of CO_2 for sustainable development involving energy, catalysis, adsorption and chemical processing. Catal. Today 2006, 115, 2-32.

[2] Dorner, R. W.; Hardy, D. R.; Williams, F. W.; Willauer, H. D., K and Mn doped iron-based CO_2 hydrogenation catalysts: Detection of $KAlH_4$ as part of the catalyst's active phase. Appl. Catal., A 2010, 373, 112-121.

[3] Dorner, R. W.; Hardy, D. R.; Williams, F. W.; Willauer, H. D., C_2-C_5^+ olefin production from CO_2 hydrogenation using ceria modified Fe/Mn/K catalysts. Catal. Commun. 2011, 15, 88-92.

[4] Satthawong, R.; Koizumi, N.; Song, C.; Prasassarakich, P., Bimetallic Fe-Co catalysts for CO_2 hydrogenation to higher hydrocarbons. J. CO_2 Util. 2013, 3-4, 102-106.

[5] Satthawong, R.; Koizumi, N.; Song, C.; Prasassarakich, P., Light olefin synthesis from CO_2 hydrogenation over K-promoted Fe-Co bimetallic catalysts. Catal. Today 2015, 251, 34-40.

[6] Wang, W.; Jiang, X.; Wang, X.; Song, C., Fe-Cu Bimetallic Catalysts for Selective CO_2 Hydrogenation to Olefin-rich C_2^+ Hydrocarbons. Ind. Eng. Chem. Res. 2018, 57, 4535−4542.

[7] Nie, X.; Wang, H.; Janik, M. J.; Guo, X.; Song, C., Computational Investigation of Fe-Cu Bimetallic Catalysts for CO_2 Hydrogenation. J. Phys. Chem. C 2016, 120, 9364-9373.

[8] Nie, X.; Wang, H.; Janik, M. J.; Chen, Y.; Guo, X.; Song, C., Mechanistic Insight into C-C Coupling over Fe-Cu Bimetallic Catalysts in CO_2 Hydrogenation. J. Phys. Chem. C 2017, 121, 13164-13174.

531878 Effect of CO_2 Source and Concentration on the Catalyst Performance and Economics of Ethane Oxidative Dehydrogenation to Ethylene

Abstract:

Southern Research, in partnership with DOE/NETL is developing a novel, low-cost mixed oxide catalyst for low temperature ethylene production process using CO_2, notably from coal-fired flue gas and readily available, low cost ethane as raw materials. The proprietary catalyst has been screened from a series of rationally synthesized catalysts and exploits CO_2 as a â€˜softâ€™ oxidant to dehydrogenate ethane to ethylene at high selectivity (oxidative dehydrogenation). The catalyst has been tested using two different CO_2 concentration levels in the feed: (a) diluted flue gas (12-14% CO_2) and (b) captured CO_2 (95%). Multiple cycle laboratory scale catalytic runs have been conducted using pure ethane and CO_2 at these two concentration levels. Each cycle constituted 7 hr of continuous run and 1 hr of regeneration. The catalyst has also been tested using several flue gas impurities, such as, SO_x, NO_x, O_2, H_2O at levels present in environmentally regulated coal fired flue gas. A techno-economic and life cycle assessment has been conducted using the laboratory scale results and considering two CO_2 concentration levels mentioned above. The economics looks competitive even with partial co-product (CO) credit. CO_2 cost was found to be one of the major drivers in the captured CO_2 case. The LCA study showed up to 57% reduction in GHG emission compared to the state-of-the-art alternative case. The presentation will highlight the impacts of CO_2 source/concentration on catalyst performance, economics and life cycle assessment.

18AIChE
Annual Meeting, Pittsburgh, PA

Utilization of CO_2 towards Solar Fuel Production via Ferrite Based Redox Reactions

(Principal Author)
Rahul Bhosale
Department of Chemical Engineering
College of Engineering
Qatar University, Doha, Qatar
rahul.bhosale@qu.edu.qa

(Co-Authors)
Gorakshnath Takalkar
Department of Chemical Engineering
College of Engineering
Qatar University, Doha, Qatar

Aliya Banu
Department of Chemical Engineering
College of Engineering
Qatar University, Doha, Qatar

Fares AlMomani
Department of Chemical Engineering
College of Engineering
Qatar University, Doha, Qatar

Anand Kumar
Department of Chemical Engineering
College of Engineering
Qatar University, Doha, Qatar

Majeda Khraisheh
Department of Chemical Engineering
College of Engineering
Qatar University, Doha, Qatar

Prepared for Presentation at
American Institute of Chemical Engineers
2018 Fall Meeting
Pittsburgh, PA, USA
October 28 – November 2, 2018

Utilization of CO_2 towards Solar Fuel Production via Ferrite Based Redox Reactions

Rahul R. Bhosale*, Gorakshnath Takalkar, Aliya Banu, Fares AlMomani, Anand Kumar, Majeda Khraisheh

Department of Chemical Engineering
College of Engineering
Qatar University, Doha, Qatar

Keywords: Ferrite, CO_2 splitting, Sol-Gel, Thermochemical

Abstract

In this investigation, a combination of cobalt and manganese based ferrite was synthesized using sol-gel method. Ethanol was used as the solvent and propylene oxide was employed as the gelation agent. The phase composition, crystallite size, specific surface area and pore volume of the derived materials was determined by using various analytical characterization techniques. In addition, the effect of calcination temperature

1. Introduction

Metal oxide based solar thermochemical splitting of H_2O and CO_2 is more feasible that the direct thermolysis due to the requirement of lower operating temperatures. The metal oxide based thermochemical splitting of H_2O and CO_2 is a cyclic process and multiple metal oxides have been tested towards these reactions [1-46]. Among these materials, cobalt based and manganese based ferrites were examined by multiple researchers and reported that these materials can successfully produce H_2 via water splitting. Although, water splitting was performed in several studies, however the CO_2 splitting ability of these materials was tested only recently and that to in very few studies. In addition, the effect of combination of both cobalt and manganese together with the ferrite as the host is not yet investigated. Hence, in this study, we have synthesized Co-Mn-ferrite ($Co_xMn_{1-x}Fe_2O_4$) for its application towards thermochemical CO_2 splitting using a thermogravimetric analyzer. The reactions associated with the $Co_{0.5}Mn_{0.5}Fe_2O_4$ based thermochemical CO_2 splitting are listed below:

$$Co_xMn_{1-x}Fe_2O_4 \rightarrow Co_xMn_{1-x}Fe_2O_{4-\delta} + {}^{\delta}/_{2}\, O_2(g) \quad (1)$$

$$Co_xMn_{1-x}Fe_2O_{4-\delta} + CO_2 \rightarrow Co_xMn_{1-x}Fe_2O_4 + CO(g) \quad (2)$$

2. Experimental

2.1 Materials

The $Co_xMn_{1-x}Fe_2O_4$ materials were synthesized using the nitrate precursors of cobalt, manganese, and iron purchased from Sigma Aldrich. The propylene oxide was used as the gelation agent and ethanol was utilized as the solvent for the sol-gel synthesis of $Co_xMn_{1-x}Fe_2O_4$ materials. Deionized water (ultrapure Type 1 from Direct –Q system, Millipore, France) was used for the preparation of the precursor solution.

2.2 Sol-gel synthesis and of $Co_xMn_{1-x}Fe_2O_4$ materials

By utilizing the propylene oxide assisted sol-gel method, the $Co_xMn_{1-x}Fe_2O_4$ materials were synthesized. First, the nitrate salts of the metal precursors were mixed together and dissolved in ethanol. After attaining the visually clear solution, propylene oxide was added dropwise and the gel formation was achieved. As-synthesized gels were aged, dried, and calcined in presence of air using a muffle furnace. The calcined powder was stored in a dry place and utilized for the characterization and thermogravimetric experiments. The process flow diagram for the sol-gel synthesis of $Co_xMn_{1-x}Fe_2O_4$ materials is presented in Figure 1.

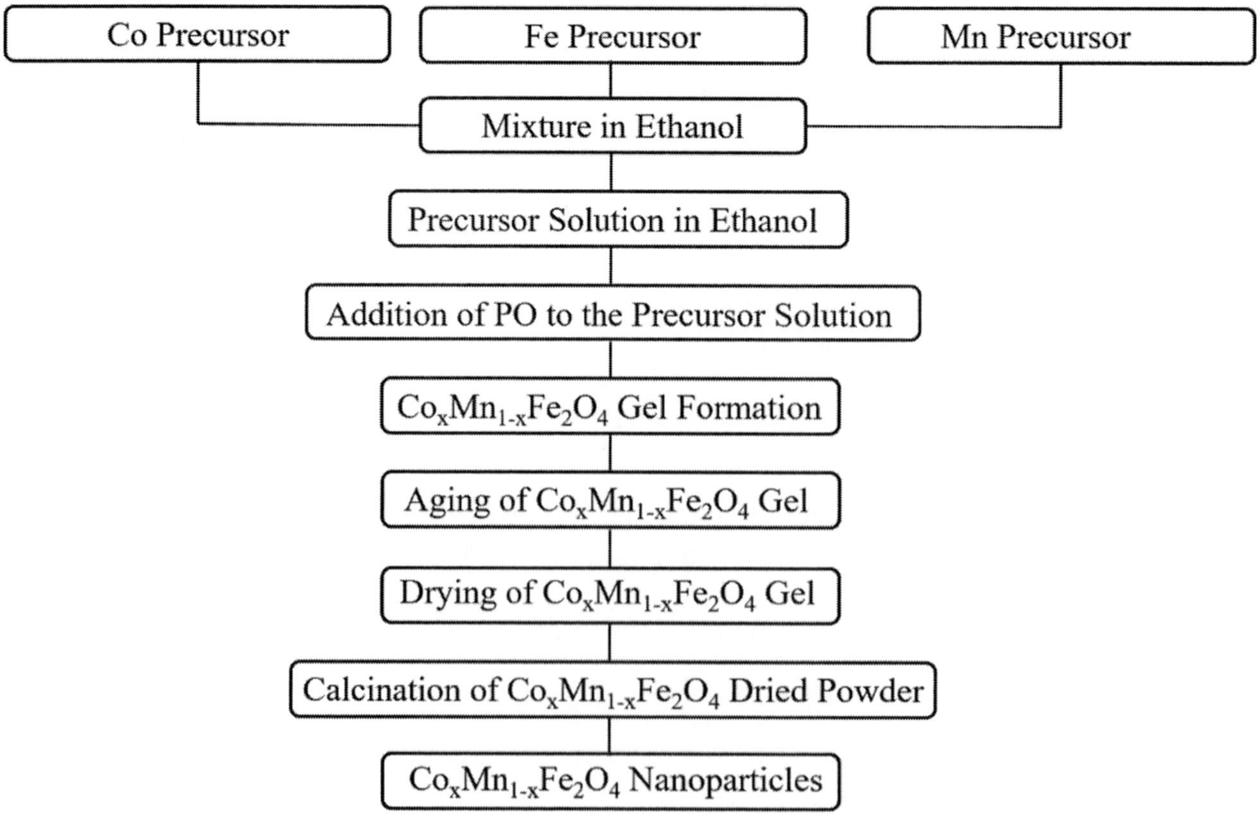

Fig. 1. Sol-gel synthesis of $Co_xMn_{1-x}Fe_2O_4$ materials (process flow diagram).

2.3 Characterization of $Co_xMn_{1-x}Fe_2O_4$ materials

The $Co_xMn_{1-x}Fe_2O_4$ materials derived via sol-gel method were characterized using powder X-ray diffractometer, BET surface area analyzer, and scanning electron microscope.

2.4 Thermogravimetric experiments

The synthesized materials are currently under testing towards thermochemical CO2 splitting using a thermogravimetric analyzer. The details about the thermogravimetric analyzer are presented in Figure 2. The $Co_xMn_{1-x}Fe_2O_4$ samples are first placed in an Al_2O_3 crucible (100 μl). The crucible-containing sample is placed on the hanging balance and multiple thermochemical cycles (containing thermal reduction and CO_2 splitting) are carried out. The variation in the mass loss and mass gain during the thermal reduction and CO2 splitting can be directly attributed to the oxygen released and CO produced. The buoyancy effect is nullified by performing blank runs (as explained in our previous publications).

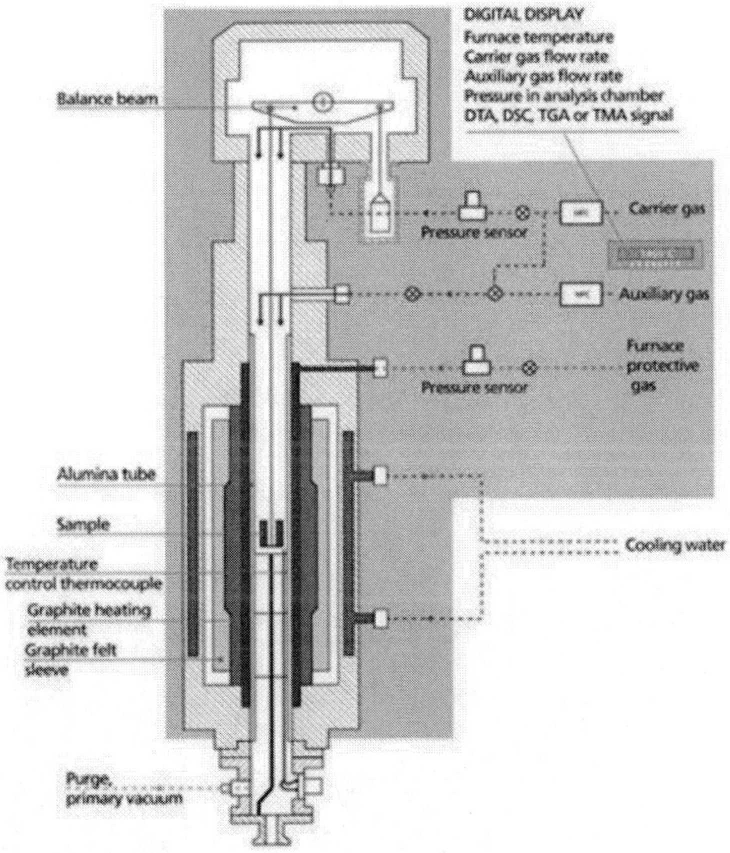

Fig. 2. Thermogravimetric analyzer useful for conducting the thermochemical CO_2 splitting.

3. Results and Discussion

The phase composition of the sol-gel derived $Co_xMn_{1-x}Fe_2O_4$ samples was determined using the powder X-ray diffractometer and the PXRD patterns obtained are shown in Figure 3. The reported PXRD patterns shows that the calcination temperature plays a very important role in the phase formation. For instance, at 600 and 800°C, a mixed phase containing $Co_xMn_{1-x}Fe_2O_4$ and iron oxide was observed. In contrast, as the calcination temperature increase up to 1000°C, a phase pure composition of the $Co_xMn_{1-x}Fe_2O_4$ was realized.

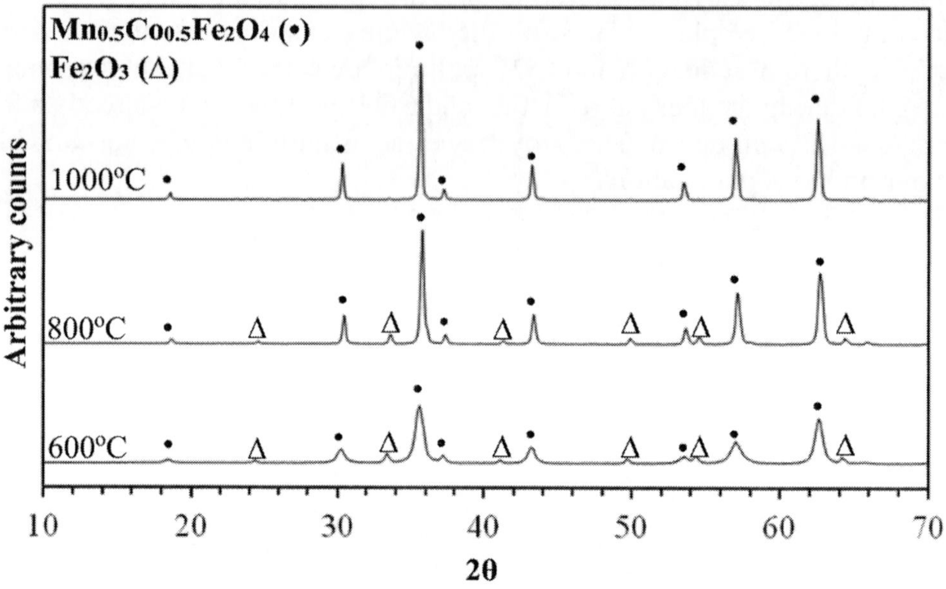

Fig. 3. PXRD patterns for the sol-gel derived $Co_xMn_{1-x}Fe_2O_4$ samples (as a function of calcination temperature).

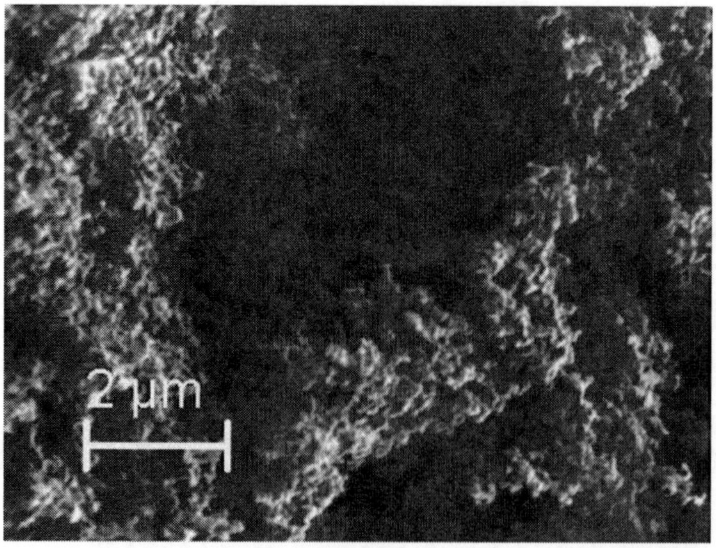

Fig. 4. SEM image for the sol-gel derived $Co_xMn_{1-x}Fe_2O_4$ nanoparticles.

The specific surface area and porosity of the derived materials were determined using a BET surface area analyzer. The specific surface area of the $Co_xMn_{1-x}Fe_2O_4$ decreases by 40% due to the increase in the calcination temperature from 600 to 800°C and by 53% when the calcination temperature was increases from 800 to 1000°C, respectively. The pore volume was also observed to be decreased by 88% whereas the crystallite size increased by a factor of 1.95 due to the rise in the calcination temperature (800 to 1000°C). The SEM image reported in Figure 4 confirms the formation of the $Co_xMn_{1-x}Fe_2O_4$ nanoparticles. The thermogravimetric CO2 splitting thermochemical experiments are under progress.

4. Conclusions

The $Co_xMn_{1-x}Fe_2O_4$ were successfully synthesized by employing the propylene oxide assisted sol-gel method. Obtained results indicate that the calcination temperature has a significant role towards the physico-chemical properties of the sol-gel derived $Co_xMn_{1-x}Fe_2O_4$.

5. Acknowledgments

This publication was made possible by the NPRP grant (NPRP8-370-2–154) from the Qatar National Research Fund (a member of Qatar Foundation). The statements made herein are solely the responsibility of author(s).

References

[1] J.R. Scheffe, A. Steinfeld, Oxygen exchange materials for solar thermochemical splitting of H_2O and CO_2: a review, Materials Today 17 (2014) 341–348.
[2] R. R. Bhosale, G. D. Takalkar, P. N. Sutar, A. Kumar, F. AlMomani, M. Khraisheh, A decade of ceria based solar thermochemical H_2O/CO_2 splitting cycle, International Journal of Hydrogen Energy (2018) Article In-Press.
[3] D. Yadav, R. Banerjee, A review of solar thermochemical processes, Renewable and Sustainable Energy Reviews 54 (2016) 497 – 532.[4] R. R. Bhosale, A. Kumar, F. AlMomani, R. B. Gupta, Solar thermochemical $ZnO/ZnSO_4$ water splitting cycle for hydrogen production, International Journal of Hydrogen Energy 43 (2017) 23474-23483.
[5] P.G. Loutzenhiser, A. Steinfeld, Solar syngas production from CO2 and H2O in a two-step thermochemical cycle via Zn/ZnO redox reactions: thermodynamic cycle analysis, International Journal of Hydrogen Energy 36 (2011) 12141–12147.
[6] R. R. Bhosale, A. Kumar, P. Sutar, Thermodynamic analysis of solar driven SnO_2/SnO based thermochemical water splitting cycle, Energy Conversion & Management 135 (2017) 226-235.
[7] A. Stamatiou, P. Loutzenhiser, A. Steinfeld, Solar syngas production from H_2O and CO_2 via two-step thermochemical cycles based on Zn/ZnO and FeO/Fe_3O_4 redox reactions: Kinetic analysis, Energy & Fuels 24 (2010) 2716-2722.

[8] R.R. Bhosale, A. Kumar, F. AlMomani, Solar Thermochemical Hydrogen Production via Terbium Oxide Based Redox Reactions, International Journal of Photoenergy 2016 (2016) 9727895.

[9] P. Loutzenhiser, M. Gálvez, I. Hischier, A. Stamatiou, A. Frei, A. Steinfeld, CO_2 splitting via two-step solar thermochemical cycles with Zn/ZnO and FeO/Fe_3O_4 redox reactions II: Kinetic analysis, Energy & Fuels 23 (2009) 2832-2839.

[10] R. Bhosale, R. Shende, J. Puszynski, Sol-gel derived $NiFe_2O_4$ modified with ZrO_2 for hydrogen generation from solar thermochemical water-splitting reaction, Materials Research Society Symposium Proceedings 1387 (2011) 34 – 39.

[11] P. Charvin, S. Abanades, F. Lemont, G. Flamant, Experimental study of SnO2/SnO/Sn thermochemical systems for solar production of hydrogen, AIChE Journal 54 (2008) 2759-2767.

[12] R. V. Shende, J. Puszynski, M. Opoku, R. Bhosale, Synthesis of novel ferrite foam material for water-splitting application, Proceedings of the 2009 NSTI Nanotechnology Conference and Expo., 1 (2009) 201 – 204.

[13] C.C. Agrafiotis, C. Pagkoura, A. Zygogianni, G. Karagiannakis, M. Kostoglou, A.G. Konstandopoulos, Hydrogen production via solar-aided water splitting thermochemical cycles: Combustion synthesis and preliminary evaluation of spinel redox-pair materials, International Journal Hydrogen Energy 37 (2012) 8964-8980.

[14] R.R. Bhosale, A. Kumar, F. AlMomani, U. Ghosh, M. S. Anis, K. Kakosimos, R. V. Shende, M. A. Rosen, Solar hydrogen production via a samarium oxide-based thermochemical water splitting cycle, Energies 9 (2016).

[15] M. Roeb, N. Gathmann, M. Neises, C. Sattler, R. Pitz-Paal, Thermodynamic analysis of two-step solar water splitting with mixed iron oxides, International Journal of Energy Research 33 (2009) 893-902.

[16] R. R. Bhosale, A. Kumar, F. AlMomani, U. Ghosh, P. Sutar, G. Takalkar, A. Ashok, I. Alxneit, Effectiveness of Ni incorporation in iron oxide crystal structure towards thermochemical CO_2 splitting reaction, Ceramics International 43 (2017) 5150-5155.

[17] J.R. Scheffe, J. Li, A.W. Weimer, A spinel ferrite/hercynite water-splitting redox cycle, International Journal of Hydrogen Energy 35 (2010) 3333-3340.

[18] R. R. Bhosale, A. Kumar, F. AlMomani, U. Ghosh, D. Dardor, Z. Bouabidi, M. Ali, S. Yousefi, A. AlNouss, M. S. Anis, M. H. Usmani, M. H. Ali, R. S. Azzam, A. Banu, Solar co-production of samarium and syngas via methanothermal reduction of samarium sesquioxide, Energy Conversion and Management 112 (2016) 413 – 422.

[19] J.R. Scheffe, A. Steinfeld, Thermodynamic analysis of cerium-based oxides for solar thermochemical fuel production, Energy & Fuels 26 (2012) 1928-1936.

[20] R. R. Bhosale, R. V. Shende, J. Puszynski, Sol-gel synthesis of ferrite foam materials for H2 generation from water-splitting reaction, Proceedings of the 2010 NSTI Nanotechnology Conference and Expo., 1 (2010) 368 – 371.

[21] W.C. Chueh, C. Falter, M. Abbott, D. Scipio, P. Furler, S.M. Haile, A. Steinfeld, High-flux solar-driven thermochemical dissociation of CO_2 and H_2O using nonstoichiometric ceria, Science 330 (2010) 1797-1801.

[22] R. R. Bhosale, R. V. Shende, J. Puszynski, Synthesis of nano-sized ferrite materials for H_2 generation from high temperature water-splitting reaction, Proceedings of the 2011 NSTI Nanotechnology Conference and Expo., 1 (2010) 237 – 240.

[23] J.R. Scheffe, R. Jacot, G.R. Patzke, A. Steinfeld, Synthesis, Characterization, and Thermochemical Redox Performance of Hf+4 , Zr+4 , and Sc+3 Doped Ceria for Splitting CO2, The Journal of Physical Chemistry C 117 (2013) 24104-24114.

[24] R. R. Bhosale, R. V. Shende, J. Puszynski, H2 generation from thermochemical water-splitting using sol-gel synthesized ferrites (MxFeyOz, M = Ni, Zn, Sn, Co, Mn, Ce), Proceedings of the 2010 AIChE Annual Meeting (2010).

[25] J.R. Scheffe, D. Weibel, A. Steinfeld, Lanthanum–strontium–manganese perovskites as redox materials for solar thermochemical splitting of H_2O and CO_2, Energy & Fuels 27 (2013) 4250-4257.

[26] R. R. Bhosale, R. V. Shende, M. K. Opoku, J. Puszynski, H2 generation from water-splitting using novel ferrite foam materials, Proceedings of the 2009 AIChE Annual Meeting (2009).

[27] A.H. McDaniel, E.C. Miller, D. Arifin, A. Ambrosini, E.N. Coker, R. O'Hayre, W.C. Chueh, J. Tong, Sr-and Mn-doped $LaAlO_{3-\delta}$ for solar thermochemical H_2 and CO production, Energy & Environmental Science 6 (2013) 2424-2428.

[28] R. R. Bhosale, A. Kumar, A. Ashok, P. Sutar, G. D. Takalkar, M. Khraisheh, F. AlMomani, U. Ghosh, La-Based Perovskites as Oxygen-Exchange Redox Materials for Solar Syngas Production, MRS Advances 2 (2017) 3365-3370.

[29] M. Gálvez, R. Jacot, J. Scheffe, T. Cooper, G. Patzke, A. Steinfeld, Physico-chemical changes in Ca, Sr and Al-doped La–Mn–O perovskites upon thermochemical splitting of CO_2 via redox cycling, Physical Chemistry Chemical Physics 17 (2015) 6629-6634.

[30] R. R. Bhosale, P. Sutar, A. Kumar, F. AlMomani, M. H. Ali, U. Ghosh, S. Al-Muhtaseb, M. Khraisheh, Journal of Renewable and Sustainable Energy 8 (2016) 034702..

[31] F. A. Costa Oliveira, M. A. Barreiros, S. Abanades, A. P. F. Caetano, R. M. Novais, R. C. Pullar, Solar thermochemical CO2 splitting using cork-templated ceria ecoceramics, Journal of CO_2 Utilization 26 (2018) 552-563.

[32] R.R. Bhosale, A. Kumar, F. AlMomani, U. Ghosh, S. Al-Muhtaseb, R. Gupta, I. Alxneit, Assessment of $Ce_xZr_yHf_zO_2$ based oxides as potential solar thermochemical CO_2 splitting materials, Ceramics International 42 (2016) 9354-9362.

[33] M. Hoes, C. Muhich, R. Jacot, G. R. Patcke, A. Steinfeld, Thermodynamics of paired charge-compensating doped ceria with superior redox performance for solar thermochemical splitting of H_2O and CO_2, Journal of Material Chemistry A 5 (2017) 19476-19484.

[34] R.R. Bhosale, A. Kumar, F. AlMomani, I. Alxneit, Sol–gel derived CeO_2–Fe_2O_3 nanoparticles: Synthesis, characterization and solar thermochemical application, Ceramics International 42 (2016) 6728-6737.

[35] C. N. R. Rao, S. Dey, Generation of H_2 and CO by solar thermochemical splitting of H_2O and CO_2 by employing metal oxides, Journal of Solid State Chemistry 242 (2016) 107-115.
[36] R. R. Bhosale. A. Kumar, F. AlMomani, I. Alxneit, Propylene oxide assisted sol–gel synthesis of zinc ferrite nanoparticles for solar fuel production, Ceramics International 42 (2016) 2431-2438.
[37] Y. Hao, C. Yang, S. Haile, Ceria–Zirconia Solid Solutions ($Ce_{1-x}Zr_xO_{2-\delta}$, $x \leq 0.2$) for Solar Thermochemical Water Splitting: A Thermodynamic Study, Chemistry of Materials 26 (2014) 6073-6082.
[38] R. R. Bhosale. A. Kumar, F. AlMomani, U. Ghosh, M. Khraisheh, A comparative thermodynamic analysis of samarium and erbium oxide based solar thermochemical water splitting cycle, International Journal of Hydrogen Energy 42 (2017) 23416-23426.
[39] T. Cooper. J. Scheffe, M. E. Galvex, R. Jacot, G. Patzke, A. Steinfeld, Lanthanum Manganite Perovskites with Ca/Sr A-site and Al B-site Doping as Effective Oxygen Exchange Materials for Solar Thermochemical Fuel Production, Energy Technology 3 (2015) 1130 – 1142.
[40] R. R. Bhosale. Thermodynamic analysis of Ni-ferrite based solar thermochemical H2O splitting cycle for H_2 production, International Journal of Hydrogen Energy (2018) Article In-Press.
[41] C. Muhich, A. Steinfeld, Principles of doping ceria for the solar thermochemical redox splitting of H_2O and CO_2, Journal of Materials Chemistry A 5 (2017) 15578-15590.
[42] R. R. Bhosale, G. D. Takalkar, Nanostructured co-precipitated $Ce_{0.9}Ln_{0.1}O_2$ (Ln = La, Pr, Sm, Nd, Gd, Tb, Dy, or Er) for thermochemical conversion of CO_2, Ceramics International 44 (2018) 16688–16697.
[43] D. Arifin, A. W. Weimer, Kinetics and mechanism of solar-thermochemical H_2 and CO production by oxidation of reduced CeO_2, Solar Energy 160 (2018) 178-185.
[44] R. R. Bhosale, Synthesis and characterization of nanocrystalline $CoFe_2O_4$ – zirconia via propylene oxide aided sol-gel method, Ceramics International 44 (2018) 191-197.
[45] R. R. Bhosale, Thermodynamic efficiency analysis of zinc oxide based solar driven thermochemical H_2O splitting cycle: Effect of partial pressure of O_2, thermal reduction and H_2O splitting temperatures, International Journal of Hydrogen Energy 43 (2018) 14915–14924.
[46] R. R. Bhosale, Solar driven two-step CH_4 reforming and H_2O splitting using Al_2O_3 for Co-production of Al, syngas, and H_2, Solar Energy (2018) Article In-Press.

533401 Efficient Selection of Conventional and Phase-Change CO_2 Capture Solvents Based on Nominal and Off-Design Process Operation

Abstract:

Solvent-based absorption/desorption is a promising and widely investigated technology for post-combustion CO_2 capture. Research efforts largely focus on the identification and selection of solvents that significantly reduce the regeneration energy requirements; the latter is the main reason for the very slow commercialization of the technology. Solvents are often evaluated and selected based on properties such as heat of absorption and loading achieved at absorption conditions. This property-based approach is useful, but it only provides indirect links to the expected solvent performance in the process where it is utilized. On the other hand, process-based solvent evaluation is a desirable approach in order to obtain a realistic evaluation of the economic and operating solvent performance. However, it is mostly performed through rigorous process and flowsheet models which provide a very detailed account of phenomena and process characteristics. The very non-ideal behavior of solvent-water-CO_2 mixtures makes the development of such reactive separation process models difficult, while their use is not without significant computational challenges. For example, numerical problems may appear due to discontinuities resulting from the appearance or disappearance of phases within a simulation. While such issues appear during simulations addressing solvent evaluation at nominal process conditions, they are amplified when off-design operation is considered to evaluate solvents at conditions other than the nominal design settings. The latter is very important for solvent evaluation because processes are essentially dynamic environments, susceptible to disturbances. A solvent that is economically desirable at nominal operation, may exhibit very high sensitivity to disturbances, resulting in the need for very expensive control approaches to maintain operation within desired set-points.

To attain process -based evaluation without numerical difficulties it is possible to use models of lower fidelity, such as shortcut models, which have been proposed in few occasions (e.g. Notz et al., 2011; Kim et al., 2015) as a means of fast, process-based solvent screening. Models such as the one proposed by Notz et al. (2011) have been shown to sufficiently capture the vapour-liquid equilibrium behavior of standard solvents (like Monoethanolamine). As they are Kremser-type models, linear approximations are required for the very non-linear equilibrium behavior of solvent-water-CO_2 mixtures which may be under- or over-estimated, while such approximations are difficult to derive. On the other hand, the model of Kim et al. (2015) focuses on the intermediate heat exchange and desorption part of the process, considering a stripper, while accounting entirely for the non-ideal behavior of the solvent in the presence of water and CO_2. The absorption operation is represented by accounting for equilibrium, whereas studying the stripper behavior permits the calculation of reboiler duty using vapor-liquid equilibrium data and energy balances around it. The approach further allows the calculation of flows, lean and rich loadings and temperatures at various points of the flowsheet, while it is only demonstrated for Monoethanolamine and Piperazine.

In this work, we adopt the model of Kim et al. (2015) and we extend and exploit it in various ways. We extend its use to additional, conventional amine solvents including solvent mixtures. We further extend the model itself to account for a new class of solvents, namely phase-change solvents. These are amines that exhibit liquid-liquid phase separation upon a change in processing conditions, e.g. upon reaction with CO_2 or upon temperature increase after reaction. The liquid-liquid phase-change enables non-thermal separation of a CO_2-lean phase prior to desorption, which is recycled to the absorber. This reduces the flow that enters the desorber, while desorption may take place at much lower temperature than with conventional solvents. As a result, it is possible to achieve very significant energetic reductions. We therefore propose a new model that is able to accommodate calculations for phase-change solvents. The model is used for evaluation of solvent performance at nominal operation, using equilibrium relations derived directly from vapour-liquid-liquid experimental data for selected phase-change solvents.

The proposed model captures directly the non-ideal solvent-water-CO_2 interactions hence it is also used to evaluate solvent performance in the presence of disturbances. This is approached through a systematic non-linear sensitivity analysis method, which investigates the static operability performance of each solvent in the process. It is based on the development of a sensitivity matrix which incorporates the derivatives of multiple process performance measures (e.g. reboiler duty, net energy penalty, cyclic capacity etc.) with respect to multiple operating parameters and solvents. The sensitivity matrix constitutes a measure of the process operating variation under the influence of infinitesimal changes imposed on the selected parameters. It is decomposed into major directions of variability associated with the eigenvectors corresponding to the larger in magnitude eigenvalues of the sensitivity matrix. The eigenvector of the largest eigenvalue represents the dominant direction of variability for the system, causing the largest change in the performance measures. The entries in the dominant eigenvector determine the major direction of variability in the multiparametric space and indicate the impact of each parameter in this direction. Having identified this direction, it is not necessary to explore all directions of variability (i.e. combinations of parameters) arbitrarily hence reducing the dimensionality of the sensitivity analysis problem. The dominant eigenvector direction is then utilised in the exploration of the system behaviour as indicated by the change of key performance indices under simultaneous, multiple and finite parameter variations.

The proposed developments are implemented for 11 solvents and mixtures, including 2 phase-change solvents. The thermodynamic equilibrium relations are derived from data available in the literature or from experiments performed internally. The process performance of few solvents is validated against results from literature sources, showing good agreement in indices like reboiler duty. The solvent-process evaluation criteria include indices such as reboiler duty, net efficiency energy penalty points, cyclic capacity, solvent mass flowrate, solvent purchase cost and lost revenue from parasitic electricity upon integration with power plants. Phase-change solvents and mixtures exhibit a reboiler duty between 2-2.3 GJ/ton CO_2, whereas the best conventional solvent mixture of 2-amino-2-methyl-1-propanol/Piperazine exhibits approximately 3.1 GJ/ton CO_2. The operability assessment highlights that certain economically desirable solvents may not be as attractive under off-design conditions.

Acknowledgements

This project has received funding from the European Union's Horizon 2020 research and innovation programme under the grant agreement 727503 - ROLINCAP – H2020-LCE-2016-2017/H2020-LCE-2016-RES-CCS-RIA.

Cited References

Kim, H., Hwang, S.J., Lee, K.S., 2015, Novel Shortcut Estimation Method for Regeneration Energy of Amine Solvents in an Absorption-Based Carbon Capture Process, Environmental Science and Technology, 49 (3), 1478–1485.

Notz, R., Tönnies, I., Mangalapally, H.P., Hoch, S., Hasse, H., 2011, A Short-Cut Method for Assessing Absorbents for Post-Combustion Carbon Dioxide Capture, International Journal of Greenhouse Gas Control, 5 (3), 413–421.

534483 Modular CO_2 Capture from Distributed Oilfield Engines

Abstract:

There are over 4,000 gas-fired rich-burn reciprocating engines (1 MW scale or higher) operating in North America, emitting an estimated 19 MMmt of CO2 per year. The majority of these engines are servicing oilfield operations and are in proximity to potential CO2 utilization opportunities. The following system design separates and captures CO2 from the exhaust gas of an engine that would otherwise be discharged to the atmosphere. Off-the-shelf components are combined in a modular design to provide a low-cost solution for capturing CO2 from small distributed sources.

This system utilizes the attractive characteristics of rich burn engine exhaust (high temperature, high CO2 concentration, low oxygen concentration) and automotive turbochargers to convert thermal energy to compression power to drive a membrane separation process.

The hot exhaust gas is routed through the shell side of a gas-gas heat exchanger where it transfers heat to a nitrogen-rich retentate stream. The exhaust is further cooled to condense and remove water and compressed from atmospheric pressure to 100-150 psia using two automotive turbochargers in series. The compressed exhaust is sent to a CO2-selective glassy polymer membrane which produces a permeate stream of 50-60% CO2 and a retentate stream of 5% CO2. The retentate stream, which is at 90-140 psia, is heated in the tube side of the gas-gas heat exchanger and then expanded in the automotive turbochargers to directly drive the compression process. The permeate stream is further concentrated using a mixture of membranes and heat-integrated refrigeration processes to either a 95% gaseous CO2 stream or a 99% liquid CO2 stream. Using turbochargers for the heat-recovery compression process significantly reduces the parasitic load demanded by an equivalent membrane process with traditional compression.

The design has been modeled in an Aspen HYSYS simulation with integrated algorithms for turbocharger and membrane performance estimations. The simulation produces performance results for various operational and ambient conditions to determine an estimation of the overall footprint, energy usage, and capital expenditure. System parameters are altered to optimize overall breakeven CO2 price.

Experimental results from a 1 MW scale turbocharger component test unit are used to validate the simulation and demonstrate the thermal energy in the rich burn engine exhaust is sufficient to drive the turbochargers. The optimized system estimates CO2 capture rate, capital expenditure, and energy consumption to produce a CO2 stream at a competitive price compared to other technologies.

538714 Application of Dynamic Reduced-Order Modeling and Advanced Process Control on UKy-CAER CO_2 Capture Pilot Plant Using CCSI Tools

Abstract:

Solvent-based processes have remained the predominant solution to post-combustion CO_2 capture for power plants. Such processes often suffer sub-optimal transient performance due to process delays involving large solvent holdups, unmeasured disturbances from solvent degradation, limited availability of measured variables, and fluctuations in ambient conditions. In this work, the 2MWth slipstream-based CO_2 capture system designed and operated by University of Kentucky's Center for Applied Energy Research (CAER) at the E.W. Brown generating station is utilized as a test-bench for developing and implementing Advanced Process Control (APC) approaches to improve transient performance and reduce energy penalty. This pilot plant involves two highly-interacting recirculation loops – a solvent loop and a desiccant loop – which result in significantly large time delays and oscillatory behavior during transient operations and serve as the major motivation for the proposed process control improvements. This work, as part of the U.S. Department of Energy's Carbon Capture Simulation for Industry Impact ($CCSI^2$) project, utilizes a suite of tools and models from the open-source Carbon Capture Simulation Initiative (CCSI) Toolset for the development of control strategies to efficiently reject process disturbances and optimize the set points within this industrial CO_2 capture process.

The CCSI's APC Framework implements Nonlinear Model Predictive Control (NMPC) and utilizes computationally efficient dynamic reduced models (D-RMs) as "fast" yet accurate predictive models. Such models are generated from the transient data measured at CAER's pilot plant by using the open-source CCSI tool D-RM Builder, which is a data-driven nonlinear system identification tool (Ma et al., 2016). Preliminary studies were conducted following certain design-of-experiments approaches and multiple sets of transient data were collected. These data were thereafter used for offline sensitivity and controllability analysis and more importantly towards developing dynamic reduced order predictive models. The developed D-RMs, with limited number of "most-influencial" input and output variables, demonstrated good prediction of plant behavior during a set of validation studies. These D-RMs were used for offline simulation-based APC studies utilizing APC Framework (Omell et al., 2016; Mahapatra et al., 2018) and suggested the potential to reduce the settling time by 60% while incurring 5% reduction in utility costs. This paper will present results from preliminary simulation-based studies utilizing the CCSI tools and discuss the development of a data-driven communication interface between APC Framework and pilot plant's distributed control system (DCS) as part of our on-going research.

Reference

1. Ma, J., Mahapatra, P., Zitney, S. E., Biegler, L. T. & Miller, D. C. (2016). D-RM Builder: A software tool for generating fast and accurate nonlinear dynamic reduced models from high-fidelity models. *Computer and Chemical Engineering*, 94, 60-74.
2. Omell, B. P., Ma, J., Mahapatra, P., Yu, M., Lee, A., Bhattacharyya, D., Zitney, S. E., Biegler, L. T. & Miller, D. C. (2016). Advanced Modeling and Control of a Solid Sorbent-Based CO2 Capture Process. *IFAC-PapersOnLine, 49*(7), 633-638.
3. Mahapatra, P., Ma, J. & Zitney, S. E. (2018). Nonlinear Model Predictive Control using Decoupled A-B Net Formulation for Carbon Capture Systems – Comparisons with Algorithmic Differentiation Approach. *American Control Conference*, Jun 27-29, Milwaukee, WI.

528195 An Optimization-Based Methodology for the Reduction of Gas Flaring in Shale Oil Production

Abstract:

Gas flaring is a common practice around the world, where nearly 5300 billion cubic feet (bcf) are burnt each year releasing 400 million tons of equivalent CO_2 emissions to the atmosphere (Farina, 2010). Governments are reacting by implementing strict environmental regulations that restrict gas flaring. For instance, in 2014 the North Dakota's Industrial Commission established a set of targets which aim to reduce flared gas to 10% by 2020 (EIA, 2014). As new regulations are implemented, several initiatives have spurred in different countries aiming at the monetization of associated gas sources that have been historically regarded as non-profitable. The World Bank is actively supporting gas flaring reduction programs through the Global Gas Flaring Reduction Initiative (GGFR) (http://www.worldbank.org/en/programs/gasflaringreduction). Nonetheless, commercialization of small gas volumes entails considerable technical and economic challenges. In this regard, great efforts are being made for developing small-scale modular plants for different processing pathways such as gas-to-liquids (GTL), gas to compressed natural gas (CNG) and liquified natural gas (LNG), and gas-to-wire (GTW). These modular plants have caught attention from different sectors due to their comparatively low capital investments and flexibility to be relocated once a gas source has been depleted. Besides gas processing, gas injection for enhanced oil recovery (EOR) is subject to extensive research as a strong alternative for reusing associated gas and increasing oil production (Sheng, 2017).

The selection of technologies for monetization of small gas volumes greatly depends on efficiencies of processing facilities, capital expenses, market conditions, and environmental aspects. Moreover, the variant nature of associated gas production adds a layer of complexity in the decision-making process. Therefore, the design of flaring reduction programs results in systems of high complexity where decisions are deeply interconnected. Optimization techniques are powerful tools that can be implemented to tackle these types of problems. This has been demonstrated in Tan and Barton (2015), where the production of liquid fuels with modular plants was addressed and shown to be profitable. Moreover, Gao and You (2017) addressed the economic and environmental feasibility of producing LNG with modular plants and conventional facilities for centralized electricity generation. Nonetheless, an integrated approach is required to exploit potential synergies between different monetization pathways.

Accordingly, this work introduces a novel optimization-based methodology that supports the design of case-specific gas flaring reduction programs via an integrated economic and environmental analysis. The novelty of the optimization framework stems from the integration of four monetization pathways: 1) physical processing monetization, 2) chemical processing monetization, 3) gas to wire monetization for distributed electricity generation, and 4) gas reinjection monetization for enhanced oil recovery. The methodology requires information such as geographic distribution and production profiles of gas sources, techno-economic information of small-scale modular plants and EOR projects, processing sites location, markets location, demand and spot prices of final products. A multiperiod Mixed Integer Linear Programming model (MILP) enables the optimization of strategic decisions such as investments plan, technologies portfolio selection, and operational decisions such as activation of well-pads, relocation of modular plants, and amount of associated gas reinjected into the reservoir. The decisions are optimized so that the environmental impact, measured as the total equivalent CO_2 emissions, is minimized and profit, measured as the net present value (NPV), is maximized. The capabilities of the proposed methodology are illustrated through a case study for the Bakken region. The case study is used to investigate economic and environmental trade-offs of associated gas commercialization in the United States.

References

EIA, 2014. Today in energy - North Dakota aims to reduce natural gas flaring [online Document]. URL https://www.eia.gov/todayinenergy/detail.php?id=18451 (accessed 12.4.18).

Farina, M., 2010. Flare Gas Reduction: Recent global trends and policy considerations, GE Energy Global Strategy and Planning.

Gao, J., You, F., 2017. Can Modular Manufacturing Be the Next Game-Changer in Shale Gas Supply Chain Design and Operations for Economic and Environmental Sustainability? ACS Sustain. Chem. Eng. 5, 10046â€"10071. https://doi.org/10.1021/acssuschemeng.7b02081

Sheng, J.J., 2017. Critical review of field EOR projects in shale and tight reservoirs. J. Pet. Sci. Eng. 159, 654â€"665. https://doi.org/10.1016/j.petrol.2017.09.022

Tan, S.H., Barton, P.I., 2015. Optimal Dynamic Allocation of Mobile Plants to Monetize Associated or Stranded Natural Gas, Part I: Bakken Shale Play Case Study. Energy, Manuscr. Rev. 93, 1581â€"1594. https://doi.org/10.1016/j.energy.2015.10.043

533847 Techno-Economic System Analysis for SOFC/GT Hybrid System Accounting for Degradation Effects

Abstract:

Techno-economic System Analysis for SOFC/GT Hybrid System Accounting for Degradation Effects

Haoxiang Lai*, Thomas A. Adams II

Solid oxide fuel cells (SOFCs) produce power with higher efficiency and lower greenhouse gas emission than conventional power production systems such as coal/natural gas power plants [1]. However, a major challenge with SOFCs is that they degrade over time, leading to a short lifetime and limiting their commercialization. When operated in constant power mode¡ªthe most common way of baseload power production¡ªthe lifetime of an SOFC is as short as around 1.5 years. As an SOFC starts to degrade, the fuel rate and current density must increase in order to compensate and keep power production at a constant level. This compounds the problem by actually increasing the rate of degradation further, resulting in an exponentially increasing degradation rate and therefore a short lifetime [2].

It has recently been found that by operating the SOFC differently with constant voltage instead of power, the degradation rate can be slowed such that the cell lifetime can be increased to around 13-14 years. In this mode, the fuel utilization decreases over time, such that the power produced by the SOFC will also decay over time to 25% of its original output at the end of the 13-14 year period [2]. In addition, the anode exhaust stream contains an ever-increasing amount of unspent fuel. Thus, this mode of operation is unsuitable for the SOFC standalone systems. In order to provide baseload power production using SOFCs with a long lifetime, one potential solution is to integrate the SOFC stack with a gas turbine (GT) in a hybrid system (as shown in Figure 1). The SOFCs operate in constant voltage mode so that their lifetime is longer (13-14 years). The GT is powered by combusting the ever-increasing unspent fuel in the SOFC anode exhaust, thus gradually increasing its power production to mostly make up for SOFC power losses incurred due to degradation over time [3]. However, the GT efficiency is lower than the SOFC stack such that it cannot make up for all of the SOFC losses. Nevertheless, the net effect is a system which can produce electricity over a long (13-14 year) lifetime with only a small amount of decay and a higher efficiency (and thus lower greenhouse gas emissions) than a standalone GT system. To contribute to the large scale adoption of SOFCs by using the SOFC/GT hybrid approach, the question that needs to be addressed is about whether the SOFC/GT hybrid system is economically feasible compared to SOFC standalone system.

Figure 1. Scheme of the Solid Oxide Fuel Cell and Gas Turbine Hybrid System.

Therefore, we present the first techno-economic study to determine if it is better to have a SOFC standalone plant in constant power mode with SOFCs replaced every 1.5 years or a more expensive SOFC/GT hybrid plant with SOFCs in constant voltage mode with SOFCs replaced every 13-14 years. Specifically, a detailed dynamic SOFC model that accounts for degradation developed in Matlab Simulink (in a prior work) was integrated with Aspen Plus steady-state models (developed in this work) of the balance-of-plant for a coal-power hybrid system. The Simulink model considers the spatial degradation over the axial length of the cells over time based on factors such as fuel composition, fuel rate, humidity, utilization, current, voltage, and other factors. The balance of plant model considers aspects such as the gasifier, syngas cleanup processes, combustor, gas turbines, compressor, and heat exchangers. The dynamics over the process lifetimes were modeled using a pseudo-steady state approach with week-long time-steps. By integrating the results of model simulation and the economic analysis, we present the trade-offs between the SOFC/GT hybrid system and the SOFC standalone system in terms of the economics and greenhouse gas emissions.

SOFC systems have been researched for over a century but they are still at their early stage of commercialization due to their high cost as being implemented in an ordinary baseload power plant. This study is valuable because it shows the techno-economic value of a SOFC/GT hybrid system that was designed to meet the needs of ordinary baseload power production with understanding of the fundamental science of how degradation occurs. This work could potentially make large scale adoption of SOFCs economically feasible.

---

[1] T. A. Adams II, J. Nease, D. Tucker, and P. I. Barton, ¡°Energy Conversion with Solid Oxide Fuel Cell Systems: A Review of Concepts and Outlooks for the Short- and Long-Term,¡± Ind. Eng. Chem. Res., vol. 52, no. 9, pp. 3089¨C3111, Mar. 2013.

[2] D. Tucker, N. F. Harun, and M. Abreu-Sepulveda, ¡°SOFC Lifetime Assessment in Gas Turbine Hybrid Power Systems,¡± p. V001T03A003, Jun. 2014.

[3] D. Tucker et al., ¡°Evaluation of Methods for Thermal Management in a Coal-Based SOFC Turbine Hybrid Through Numerical Simulation,¡± J. Fuel Cell Sci. Technol., vol. 9, no. 4, pp. 041004-041004-9, Jun. 2012.

524796 Coal Drying, Mercury Removal, and Stabilization Using Heat from a Nuscale Small Modular Reactor

Abstract:

Low rank coals tend to have high moisture levels. In our experiments, we worked with Powder River Basin subbituminous coal, which typically has 25 to 30 wt. % moisture. Drying these coals has long been recognized as a means of increasing power plant efficiency, and possibly reducing shipping costs. The heat for drying can be provided by a nuclear reactor, such as the small, modular, nuclear reactor under development by NuScale Power, which produces steam at 280 to 300 °C.

We investigated drying coal at temperatures from 100 to 350 C. We found that mercury removal increases with increasing drying temperature. About three quarters of the mercury originally contained within the coal is removed when it is dried at 350 C.

Low rank coals, when dried, are prone to low temperature oxidation and this can lead to spontaneous combustion. This is a primary barrier to wide scale commercial coal drying. We found that controlled exposure to air at 150 to 250 C eliminates the tendency of the coal to further oxidize at low temperatures. We found, however, that the dried and stabilized coal tends to re-absorb moisture from humid air, which reverses some of the benefits of drying.

538195 Optimal Design of Gas-Fired Moving-Bed Chemical Looping Combustion Systems

Abstract:

Considerable improvement in computational power and numerical algorithms over the past few decades has enabled the use of large-scale, first-principles equation-oriented process models for the design, scale-up and optimization of new and existing technologies. While several efficient commercial simulators offer a library of basic unit operations and support the development of equation-oriented mathematical models of complex unit operations and flowsheets, they often have limited capabilities for solving large-scale process system engineering optimization problems. In contrast, algebraic modeling languages (AMLs) enable the simultaneous solution and optimization of high-complexity large-scale process models, but do not provide a library of unit models, and building the required models and flowsheets often becomes a demanding, time consuming task.

The Institute for the Design of Advance Energy Systems (IDAES) is developing a comprehensive, next generation computational framework [1] to facilitate the design and optimization of innovative energy systems and related chemical engineering systems. The IDAES framework is developed in Pyomo [2], an open-source Python-based AML, and not only supports a wide range of optimization solvers, but also provides a library of open (editable) standard and complex unit operation models [3]. In this study, we demonstrate the capabilities of the IDAES framework by presenting the development of a complex unit model and its use in a multi-objective flowsheet optimization problem. The application area of interest is the optimization of methane-based chemical looping combustion (CLC) systems using different oxygen carrier materials.

CLC is a novel, promising advanced energy technology for the conversion of fossil fuels with inherent carbon capture. It consists of the cyclic reduction-oxidation of an oxygen carrier, generally a metal oxide, circulating between two reactors: a fuel (reducer) reactor and an air (oxidizer) reactor. The oxygen necessary for fuel conversion in the reducer reactor is provided by the oxygen carrier, thus ensuring combustion in pure oxygen rather than air. The exhaust gas leaving the fuel reactor is mainly a mixture of carbon dioxide (CO_2) and water (H_2O) from which H_2O can be condensed resulting a CO_2-rich stream ready for utilization or storage. Under carbon-constrained scenarios, CLC is an attractive alternative to conventional technologies due to its high efficiency and lower cost compared with post-combustion CO_2 capture.

The goal of this work is the optimal design of the chemical looping reactors for methane combustion carried out in two interconnected moving-bed (MB) reactors. Detailed models of MB reactors for CLC are currently lacking in the literature. In our previous work, a 1-D steady-state, nonisothermal, first-principles model of a MB reactor with hydrodynamics under reduction conditions was developed. The model is flexible and modular, and it can be readily adapted for the simulation of different gas-solid processes [4]. In this study, first an oxidizer reactor model is developed by modifying the reducer reactor model. Then, the development of a flowsheet for the CLC of methane using three promising oxygen carriers, Fe-, Cu-, and Ni-based carriers, is presented. Detailed property packages along with the kinetic models [5] governing the reduction and oxidation reactions of the different oxygen carriers are developed and used in the MB models. Finally, the equation-oriented process flowsheet is used to optimize the efficiency of the fuel reactor, while minimizing the capital and operational costs. The large-scale multi-objective optimization problem is solved using IPOPT. The feasibility of the interconnected MB reactor system for the combustion of methane with each oxygen carrier is investigated and the optimal design of the reactors is determined.

References

1. Miller D.C., Siirola J.D., Agarwal D.A., Burgard A.P., Lee A., Eslick J.C., Nicholson B.L., Laird C.D., Biegler L.T., Bhattacharyya D., Sahinidis N.V., Grossmann I.E., Gounaris C.E., and Gunter D., 2018, "Next Generation Multi-Scale Process Systems Engineering Framework", *Proceedings of the 13th International Symposium on Process Systems Engineering*, San Diego, California, USA.
2. Hart W.E., Laird C.D., Watson J.-P., Woodruff D.L., Hackebeil G.A., Nicholson B.L., and Siirola J.D., 2017, "Pyomo – Optimization Modeling in Python", Second Edition. Vol. 67. Springer.
3. Lee A., Ghouse J.H., Chen Q., Eslick J.C., Siirola J.D., Grossmann I.E., Miller D.C., 2018, "A Flexible Framework and Model Library for Process Simulation, Optimization and Control", *Proceedings of the 13th International Symposium on Process Systems Engineering*, San Diego, California, USA.
4. Ostace A., Lee A., Okoli C.O., Burgard A.P., Miller D.C., and Bhattacharyya D., 2018, "Mathematical Modeling of a Moving-Bed Reactor for Chemical Looping Combustion of Methane", *Proceedings of the 13th International Symposium on Process Systems Engineering*, San Diego, California, USA.
5. Abad A., Adanez J., Garcia-Labiano F., de Diego L.F., Gayan P., and Celaya J., 2007, "Mapping of the range of operational conditions for Cu-, Fe-, and Ni-based oxygen carriers in chemical-looping combustion", *Chemical Engineering Science,* 62, pp. 533-549

542620 The Impact of Shale Gas and Oil on the Chemical Industry

Abstract:

Lower alkanes and aromatics are very useful as both fuels and chemical raw materials. Occasionally supply disruptions constrained availability but also encouraged investigation of alternative feedstocks and chemistries. Recent new technologies for the production of unconventional gas and oil sources from tight shale formations significantly increased availability of both methane and condensate, radically altered process economics, resulted in the construction of much new intermediates capacity, and multiplied the ways traditional and alternative chemistries might be exploited. This presentation will outline historically how natural gas components became useful as chemical feedstocks, how government regulations occasionally led to perturbations in availability and price, how supply crises led to the development of alternative chemistries, catalysts, and ultimately the exploitation of shale resources. More specifically, will be discussed how excess coproducts, especially condensate, from increasing shale gas and oil production for energy are having significant unexpected impacts on the chemical process industries and opportunities for new process chemistry research.

541186 Sustainable Energy and Chemicals: Past, Present, and Future

Abstract:

The last 40 years has seen the U.S. energy situation vacillate from gasoline rationing and concerns over Peak Oil, to projected energy and feedstock prices that were noncompetitive in global markets for chemicals production, to the economic revolution of frack gas and light tight oil. The importance of low cost energy and feedstocks in driving economic opportunity will be discussed relative to this history, along with the importance of sustainability in decision making to gain stakeholder buy-in and acceptance. Strategies for managing career paths and competencies despite changing targets and high uncertainty will be addressed.

542618 Disruptions: What the Future May Hold

Abstract:

Disruptive forces are all around us. They change the way we communicate and work. They change the way we develop goods and services. They also change the types of products that are manufactured. Examples such as the way in which we produce and consume energy or the feedstocks we use to produce product have seen many changes over the course of the last few decades, and these disruptions are expected to continue. The chemical engineering discipline and the expectations for a chemical engineerâ€™s skills has also changed and adapted as a result of these disruptive forces. A discussion of the broad themes associated with these forces and the implications that they may have to chemical engineers in the future will be discussed.

542621 Geopolitical Factors Influencing the Evolution of the Chemical Industry

Abstract:

The basic chemicals and plastics industry is built on a simple principle: find a cheap feedstock and convert it into something more valuable. However, the right feedstock depends on place and time. Certain policies and actions of various governments have influenced the evolution and profitability of the chemical industry during the past 50 years, sometimes in surprising ways. For example, high oil prices and US government funding encouraged the development and commercialization of coal gasification technology during the late 1970s and early 1980s. This enabled China to develop a competitive basic chemical industry using coal as a feedstock during the last 25 years. Similarly, an abrupt increase in oil prices during the first decade of the 21st century encouraged the present shale gas revolution, causing a major shift in the economy of producing basic chemicals and polymers in the US. If history is our guide, then what governments choose to fund and regulate may be as important as market forces in determining the shape of our industry and profession in the future.

542622 Agility & Resilience: How to Maintain Career Competitiveness in the Changing Chemical Industry

Abstract:

The world around is changing. The shift to protectionism, global instability and price volatility of raw materials and feedstocks has the potential to wreak havoc on the chemical industry. With the revolution of shale gas in North America and the sudden drop in feedstock costs, the chemical sector in the US has reasserted its historic dominance. Learn how the role of a chemical engineer can contribute to industry growth and resilience, while adapting to allow for personal agility throughout the career life cycle.

539274 Catalytic Aromatization of Bio-Derived Liquid Under Methane Environment

Abstract:

Hydrodeoxygenation upgrading is the most popular way to produce bio-oil from biocrude. This approach usually involves continuous hydrogen flow and high operation pressure (15~100 atm), resulting in significantly increased capital and operational costs. Compared to hydrogen, which is not naturally available and thus costly, methane can be readily obtained as natural gas. This work studies the technical feasibility of directly using methane as the reducing agent instead of hydrogen to remove oxygen from a series of model compounds including ethanol, acetic acid, acetone, and phenol, representatives of biocrude derived from low cost carbon resources at low pressure (<5 MPa) and moderate temperatures (400 °C) under the facilitation of transition metal modified aluminosilicate catalysts, the produced aromatics abundant liquid will become more cost competitive. Moreover, methane itself is incorporated into the formed aromatics beneficial for extra liquid chemical production along with lower carbon dioxide (CO_2) emissions from the upgrading process. The involved catalytic upgrading mechanism is also intensively studied using isotopic labeling technique coupled with versatile characterizations including NMR, FTIR, XAS, XPS, SIMS, and DSC as well as molecular simulation at DFT level for verifying methaneâ€™s participation and benefiting catalystâ€™s rational design. The outcomes from this research will provide a more efficient and cleaner way to utilize low cost carbon resources and natural gas which are abundant in nature. Furthermore, the successful implementation of the study will be beneficial for the effective reduction of greenhouse gas emission and significant cost reduction of the current bio-oil production process making it more economically feasible.

529825 Wave Liquefaction™ Processing of Carbon Materials for the Production of Value-Added Chemicals and Feedstocks: Spectroscopic Diagnostics and Material Characterization

Abstract:

1.0 Introduction

This paper presents optical spectroscopic diagnostics as used in an engineering pilot implementation of H Quest Vanguard's microwave plasma process applied to coal, and material characterization results from application of the process to H_2 generation from natural gas. Plasma diagnostics for process control will be illustrated. Chemical analyses are presented of coal conversion products. Recent results yielding nanographene as a premium carbon by microwave-plasma processing of methane will be shown.

Microwave-driven plasma-mediated reaction engineering represents process intensification for energy efficient and economical hydrocarbon conversion processes. Examples include coal-to-liquids and natural gas conversion to hydrogen and premium carbon products. Optical emission spectroscopy and analysis provides species identification and determination of temperatures in different energy modes in the non-equilibrium plasma.

Microwave energy, variable between 1 - 4 kW is coupled into a continuous flow reactor. Wave Liquefaction™ process applied to coal implements in-situ product upgrading via hydrogenation and/or methylation reactions enabled by activating hydrogen or methane present in the process gas. Pulverized coal is fed by a particle seeder. Gas ratios are optimized for the particular hydrocarbon feedstock conversion. An alternative application of the microwave plasma processing is methane decomposition to H_2 and premium carbon products. Reactor residence times are sub-second and performed at atmospheric pressure. Filters, traps and cyclone stages downstream of the reactor perform gas/liquid/solid separations.

2.0 Optical Emission Spectroscopy

Optical diagnostics are central to reaction characterization and hold particular value for species identification and temperature determination [1]. Observed intensity ratios or spectra band shapes can yield temperature by Boltzman analysis using spectral constants. Moreover, optical emission serves to identify reactive species and intermediates, albeit indirectly inferred by the observation of their electronically excited counterparts, e.g. CH* and C_2* radicals. For example, the presence of key atomic or diatomic radicals can support postulates of electron impact dissociation, and radical mediated bond insertion or radical capping reactions and provide mechanistic insights from the temperatures associated with the different degrees of freedom – electronic, vibrational, such as from the C_2^* ($d^3{}_g - a^3{}_u$) Swan band emission. In this study spectroscopic data collection was performed using a fiber-optically coupled Ocean Optics HR2000 spectrometer fitted with a UV-NIR grating for a spectral range of 900 nm. All optical emission spectra were captured with equipment spectrally calibrated using a Hg/Ar lamp and corrected for instrument response function using a NIST-traceable tungsten lamp as spectral calibration standard. Temperatures of C_2^* are determined using custom spectral band-fitting algorithms, taking into account the electronic baseline and underlying blackbody radiation. Spectral fits are performed at 1 nm spectral resolution. Carbon particle temperatures are determined by fitting Planck's radiation law with black body conditions. Ar* emission lines are identified by reference to the NIST spectroscopic database.

3.0 Gas/Liquid/Carbon Analyses

Microwave (MW) activated natural gas decomposition for H_2 generation produces no CO_2 and requires zero process water, in contrast to the present industrial process – steam reforming. Notably it opens a path for renewable energy storage by coupling electrical energy into chemical bond energy – with H_2 being particularly versatile as fuel or chemical precursor. Moreover, MW activated natural gas decomposition does not require additional carbon as catalyst but rather produces carbon [2]. Therein MW driven decomposition of methane achieves green hydrogen production and production of value-added carbon materials [3]. Here a suite of characterization tools have been applied to the light gases, liquid and solid carbon products. Gas and liquid analyses were performed using gas chromatography-mass spectrometry. For the solid carbon products SEM has been applied to coal chars for cenosphere analysis, and for the carbon products from methane decomposition, scanning electron microsocpy (SEM) for morphology, transmission electron microscopy (TEM) for (nano)structural assessment and identification of different carbon forms (i.e. sp^2 phases), X-ray diffraction (XRD) for evaluation of graphitic structure, thermo-gravimetric analysis (TGA) for bulk determination of oxidative reactivity of the carbons as a means by which to assess graphitic content and gas (N_2) adsorption analyses for texture – i.e. surface area and porosity. Elemental analysis was also performed for HCNS and O by difference. Complementary thermo-gravimetric analyses assess the oxidative reactivity of the carbon products. Optical diagnostics such as multi-wavelength pyrometry relate these material characteristics to reactor conditions and can be applied for process control.

4.0 Summary

H Quest Vanguard is developing broad-spectrum microwave plasma processes targeting conversion of hydrocarbon feedstocks such as coal and natural gas to value-added materials, chemicals and fuels [4,5].

A microwave pyrolysis approach was originally proposed in response to the DARPA initiative for green-house gas (GHG)-emission-free and cost-effective production of US Air Force jet fuel from the domestic coal resources. In H Quest Vanguard's process, natural gas can be used in single-stage reactor as a hydrogen source, eliminating external hydrogen production units and the associated CO_2 production, water consumption, and capital costs, and providing excess hydrogen sufficient for downstream hydro-treating. The MW process produces fast heating resulting in flash devolatilization and pyrolysis followed by fast quenching, preserving the primary pyrolysis products and volatiles' molecular structure.

Since company formation in 2014, H Quest Vanguard, Inc. has developed ample material base and expertise in development of chemical and catalytic processes enhanced by microwave plasma. In particular, H Quest has developed a conversion process that applies microwave energy to rapidly co-pyrolyze solid hydrocarbons (e.g. coal) and natural gas to produce liquid hydrocarbons. A wide range of carbon materials including graphene and ordered carbon blacks have been observed across a wide range of experiments. These forms have potential for high value applications: electrical conductivity additives for plastics, and as electrode material in supercapacitors and Li-batteries. Subsequent and ongoing work has demonstrated feasibility of a microwave driven plasma mediated methane decompostion for H_2 production and nanographene formation as a premium carbon.

5.0 Acknowledgements

H Quest Vanguard, Inc. is a privately held technology company, based in Pittsburgh, Pennsylvania, focused on the development and commercialization of novel hydrocarbon conversion technologies. This material is based on work supported by the Department of Energy, Office of Science through subaward agreement no. 186949 with H Quest Vanguard, Inc. under the Prime Award DE-SC0015895 Phase I SBIR.

6.0 References

[1] Vander Wal, R. L., Gaddam, C. K., and Kulis, M. J. (2014). An Investigation of Micro-Hollow Cathode Glow Discharge Generated Optical Emission Spectroscopy for Hydrocarbon Detection and Differentiation. *Applied spectroscopy, 68*(6), 649-656.

[2] Dagle, R., Dagle, V., Bearden, M., Holladay, J., Krause, T., and Ahmed, S., (2017). *An Overview of Natural Gas Conversion Technologies for Co-Production of Hydrogen and Value-Added Solid Carbon Products* (No. PNNL-26726; ANL-17/11). Pacific Northwest National Lab. (PNNL), Richland, WA (United States); Argonne National Lab. (ANL), Argonne, IL (United States).

[3] Abbas, H. F., and Daud, W. W. (2010). Hydrogen production by methane decomposition: A review. *International Journal of Hydrogen Energy,* 35(3), 1160-1190.

[4] Skoptsov, George L., and Alan A. Johnson. *Method for processing hydrocarbon fuels using microwave* energy. U.S. Patent 9,682,359, issued June 20, 2017.

[5] Strohm, J.J., Linehan, J.C., Roberts, B.Q., McMakin, D.L., Sheen, D.M., Griffin, J.W. and Franz, J.A., Battelle Memorial Institute, 2012. *Heavy Fossil Hydrocarbon Conversion and Upgrading Using Radio-Frequency or Microwave Energy*. U.S. Patent Application 13/401,216.

537027 Hydrocarbon Chain Growth Via a Nonthermal Electrical Plasma Microreactor

Abstract:

Currently, large amounts of methane are emitted or flared to the environment from stranded sources such as landfills and agricultural sources. The infrastructure required to bring these resources to market is cost prohibitive. There is an economic opportunity to convert this methane into a more transportable or value-added product that can be brought to market.

Using a nonthermal electrical plasma, reactions that usually require high temperatures and pressures can actually take place close to ambient conditions. Instead of thermal collisions of molecules causing reactions, high energy electrons collide with molecules in order to disrupt bonds. In this case, carbon-hydrogen bonds are broken which allows for longer hydrocarbons to form. This process therefore transforms electrical energy into chemical energy.

Performing these reactions at the microscale level has the advantage of decreasing the space between the electrodes that create the electrical plasma. By reducing the size, less voltage is required in order to generate these plasmas which reduces the costs of the necessary power electronics.

Recent results of the chemical species produced by the process and the energy efficiency of the plasma will be discussed.

525092 Combined Experimental and Density Functional Theory Studies on the Modified $SrTiO_3$ Catalysts for Oxidative Coupling of Methane

Abstract:

The main component of natural and shale gas, methane, attracted global attention because of the richness of deposits and significant potential for direct or indirect conversion into value added chemicals and fuels. Oxidative Coupling of Methane (OCM) is one of promising methods of adding value to methane via directly producing ethane and ethylene, vital building blocks in chemical industry. We studied modified $SrTiO_3$ based perovskites with metal dopants which were prepared by polymeric precursor method. $A_xSr_{1-x}TiO_3$ (A = Ba, Ca, La, Y, Li, Na, K, Rb, Cs, Al, Mg, Tl, and Pb, x = 0.2) and $SrB_xTi_{1-x}O_3$ (B = Sn, Al, Mg, Ce, Pr, Nd, Sm, and In, x = 0.2) were investigated. Also, we use density functional theory (DFT) calculations to remove ambiguity on the identification of the catalytically active site and devise descriptors for high-performance catalyst. The volcano curve was observed for C_{2+} selectivity with methyl radical adsorption energy from experimental and DFT results. The correlation indicated that the appropriate strength of methyl radical adsorption is required for improving C_{2+} compounds formation. Based on catalyst characterization using X-ray diffraction (XRD), O_2-pulse injection and temperature programmed desorption (O_2-TPD), and X-ray photoelectron spectroscopy (XPS), the influences on catalytic activity by partial replacement of cations with A- and B-site dopants were investigated.

538208 Process Technology Assessment for Mega Scale Projects

Abstract:

The assessment of process technologies is an important part of integrated process engineering for enterprises to make strategic decisions on new commercialization investment or on the expansion of existing operations with cutting-edge technologies. It becomes even more critical especially when a project requires technologies that are unfamiliar to an organization.

This paper discusses a basic understanding of the work required and the fundamentals in evaluating innovative process technologies for mega scale projects in chemicals manufacturing. The assessment principles, criteria and the processes will be discussed in both technology, economics as well as sustainability perspectives. Coal Gasification technology for Syngas manufacturing is used as an example.

528317 Catalyst Development for Natural Gas Monetization

Abstract:

Recent abundance of natural gas rejuvenated interest in catalytic methods leading to its high value products. Developments in catalyst design and structural details under working conditions for sulfur and carbon containing natural gas components, such as CH_4, CO_2, H_2S and ethane, will be presented. First, we will focus on the molecular structure and properties of Oxidative Coupling of Methane (OCM) catalyst. Emphasis will be on using spectroscopy tools to interrogate the active site of OCM supported tungstate catalyst with the goal of later controllably synthesizing it. Second, both CO_2 and H_2S acid gases are present in significant concentrations in natural â€œsourâ€ gas streams and need to be separated using energy intensive amine absorption methods. The captured CO_2 is emitted into the atmosphere acting as a major greenhouse gas while H_2S is combusted to yield elemental sulfur and water. A catalytic method of deriving higher hydrocarbons from CO_2 and H_2S under reducing environment will be presented and the fundamental knowledge base of acid gas - CO_2 and H_2S as well as they major reaction byproduct SO_2 - induced K_2O-WO_3/Al_2O_3 catalyst structural changes will be investigated using *in situ* Raman, IR spectroscopy and TPSR. Finally, solid acid catalyst design and molecular structure details will be presented for ethylene coupling reactions.

538680 Catalytic Wall Reactor for Non-Oxidative Methane Conversion

Abstract:

Methane, if converted to higher hydrocarbons, promises a great future as the substituent for liquid petroleum in petrochemical and fine chemical industries. Methane conversion methods such as oxidative coupling of methane (OCM) and non-oxidative methane aromatization (DMA) reactions have attracted much attention since they directly convert methane into useful hydrocarbons including C_2 (ethylene and ethane) and benzene. In the OCM process, the more reactive nature of C_2 products than methane leads to the sequential oxidation of C_2 to thermodynamically favored CO_x (CO or CO_2). Similarly, the DMA process is challenged by the low methane equilibrium conversion and rapid catalyst deactivation. In this talk, we will highlight our recent work on development of catalytic wall reactor for non-oxidative methane conversion into C2 , aromatics and hydrogen products. The reaction temperature and gas flow rate were tested as variables to tailor methane conversion and product selectivity. The energy and economic analyses were conducted on the developed reactor systems to evaluate their practical application feasibility.

537324 Exploring Strategies to Improve Yields of Oxidative Coupling of Methane in a Chemical Looping System

Abstract:

Direct conversion of methane to value added products such as hydrocarbons has been a grand challenge for many decades. Although there are several strategies and technologies for converting methane, oxidative coupling of methane (OCM) is a promising pathway which is the focus of this study. Under chemical looping mode, methane is oxidized by a metal oxide, instead of molecular oxygen as done in the traditional co-feed OCM to produce higher hydrocarbons as the desired product. The metal oxide thus serves as a catalytic oxygen carrier (COC), which provides lattice oxygen to oxidize hydrogen and oxygen vacancies to serve as active sites for methyl radical formation and desorption. Previously, a detailed energy analysis was performed on the chemical looping OCM process over a proprietary Mg_6MnO_8 based COC, which gave a hydrocarbon yield of 23.2%. The objective of this study is to intuitively develop COCs which can achieve high yields in OCM by modifying the base COC.

This study would focus on using density functional theory (DFT) calculations to screen for different metal oxide dopants which are potential candidates that can increase the OCM hydrocarbon yield. This computational tool will be accompanied by experimental screening to develop an optimal COC configuration. Thermogravimetric analysis (TGA) will investigate the reactivity of these COCâ€™s under methane reduction. TGA would also help in investigating the recyclability of the COC over several redox cycles. These results will be combined with fixed bed tests of the COCs, which would provide gas-composition data as the reduction proceeds. A fourier transform infrared (FTIR) gas analyzer along with a gas chromatography analyzer will be used for this purpose. Changes in the COC will be characterized by using X-ray diffraction (XRD), X-ray photoelectron spectroscopy (XPS) and Raman spectroscopy. These results will be fit into a microkinetic model, which would be developed from the experimental and theoretical calculations. Results from this study would provide key insights into the development of such COCs for chemical looping application.

527336 Metal-Mediated Transient Hydrogen Scavenging for Enhanced Aromatics Yield during Non-Oxidative Methane Aromatization on Mo/H-ZSM-5 Catalysts

Abstract:

We report a polyfunctional catalyst formulation that comprises a hydrogen-selective zirconium adsorbent with a Mo/H-ZSM-5 methane dehydroaromatization catalyst that results in enhanced methane conversion, accelerated rates, and increased selectivity to desired products. Chemical transient experiments show that 2.4 ± 0.1 O:Mo are lost as CO, CO_2 and H_2O during initial CH_4 reactions with MoO_x precursors in a stoichiometric reaction that leads to MoC_x evincing the formation of $(Mo_2O_5)^{2+}$ dimers on air treatment of MoO_3/H-ZSM-5 physical mixtures at 973 K. This synthetic protocol gives a Mo/H-ZSM-5 formulation with a steady-state net benzene formation rate of $2.5x10^{-4}$ mol mol_{Mo}^{-1} s^{-1}. Mo/H-ZSM-5 produces ethene, benzene, and naphthalene with high selectivity via bifunctional catalysis where MoC_x species catalyze the C-H bond activation of methane to ethene/ethane, residual zeolitic BrÃ¸nsted acid sites catalyze olefin coupling for chain growth, and MoC_x species catalyze aromatic formation through aliphatic dehydrogenation. Ethylene, benzene, and naphthalene yields on Mo/H-ZSM-5 approach values prescribed by equilibrium ($6CH_4$ â†” $C_6H_6 + 9H_2$; equilibrium conversion ~10% at 973 K). Addition of zirconium (Zr) particles to the MoC_x/ZSM-5 formulation results in methane conversion of ~30% demonstrating that the limitation of equilibrium conversion of ~10% at 973 K is circumvented due to hydrogen scavenging by formation of a hydride species on the adsorbent Zr. A 3.1 times increase in methane converted with a 1.4, 2.1, 2.6, and 5.6 times increase in benzene, naphthalene, toluene, and C_{10}^+ yields respectively (after 8.52 ks time-on-stream) are achieved via interpellet mixtures of Zr and MoC_x/ZSM-5 as compared to conventional MoC_x/ZSM-5 catalyst. Furthermore, ~8.5 mol % of methane converted is deposited as coke at the end of 8.52 ks time-on-stream. Subsequent thermal treatment of the catalyst results in desorption of hydrogen and in regeneration of the Zr adsorbent leading to partial regeneration of the polyfunctional catalyst formulation.

We discuss the transients in rates and selectivity for the Zr + MoC_x/ZSM-5 formulation as a result of spatial gradients in hydrogen concentration across the reactor length due to the gradual, stoichiometric reduction of the metal adsorbent to form a hydride as well as loss of acid sites due to coke deposition. Detailed catalyst characterization, hydrogen uptake experiments on Zr adsorbent, and investigations of relevant length scales will also be presented.

525600 Promotional Effect of Cr in Mo_2CÂ catalyst Supported on Sulfated Zirconia for Methane Dehydroaromatization

Abstract:

Promotional effect of Cr in Mo_2C catalyst supported on sulfated zirconia for methane dehydroaromatization

Ashraf Abedin[1], Swarom Kanitkar[1], Srikar Bhattar[1], James J. Spivey*[1]

[1]Cain Department of Chemical Engineering, Louisiana State University, Baton Rouge, LA, USA

The need for valuable hydrocarbons in chemical engineering sectors is now at its peak. Methane, having an abundant source of supply from natural gas, has been well-studied as feedstock for producing heavier hydrocarbons of high value. A number of methane activation processes have been developed. Methane dehydroaromatization (MDHA) is a direct methane conversion process, eliminating the need for a syngas step while producing both benzene and hydrogen.

The mechanism of MDHA consists of two reactions. The first step is the activation of methane C-H bonds to produce CH_x species. These free radicals react to generate C_2H_y dimers. The dimers further oligomerize to produce C_6H_6 and related homologues. In recent years, Mo oxide species supported on ZSM-5/MCM-22 has been widely investigated for MDHA. Mo oxide is carburized to produce Mo_2C, which is believed to activate the C-H bonds of methane to C_2H_y dimers. The acidic sites on ZSM-5/MCM-22 oligomerize the dimers to produce benzene and higher homologues. [1]

In this work different loadings of Cr were used as promoters for the Mo oxide based bifunctional catalyst. Cr has been previously added along with other metals in ZSM-5 in the form of extra-framework species to facilitate the catalyst dehydrogenation function for MDHA. [2] Cr_3C_2 species has electron configuration similar to the active Mo_2C at the Fermi level. This property matches with that of noble metals like Pt [3], which showed high activity for MDHA [4]. The hypothesis is that the additional active sites would generate more dimers available for oligomerization. Sulfated zirconia (SZ) provides acidic sites for MDHA, and was used as a novel support replacing the conventional ZSM-5.

Three different SZ catalyst were prepared, each having 5 wt. % of Mo with Cr loading of 0, 0.12 and 0.5 wt. % respectively. Incipient wetness impregnation was used to dope the metal oxides onto SZ and the catalysts were dried overnight at 110 C, followed by calcination at 550 C for four hours. Raman spectroscopy was used to confirm the loading of the metal oxides. The surface acidity was determined by FT-IR spectra of chemisorbed pyridine.

To compare the activity of the three different catalysts, each run was made under identical reaction conditions. The catalysts were loaded in a continuous flow reactor system and reduced under H_2 flow till the reaction temperature of 700 C was reached. Following this a 1:4 CH_4:H_2 flow was maintained for four hours to achieve the carburized form. MDHA reaction was run afterwards for over 15 hours to investigate the activity of the promoted catalysts. Reaction products were analyzed using GC-FID-TCD.

Primary products include Ethylene, Ethane, Propylene, Propane and aromatics (benzene, toluene and ethylbenzene). During a 15-hr run, the 0.5 wt. % Cr promoted catalyst had the highest of methane conversion, whereas the other two catalysts showed lower and virtually identical conversions (Figure 1). The 0.5 wt. % Cr promoted catalyst showed the highest Benzene selectivity, while the 0.12 wt. % Cr promoted and the unpromoted catalyst (Figure 2) had essentially the same benzene selectivity. Based on this finding, it can be proposed that a significant amount of Cr loading can modify the catalytic surface of Mo_2C-SZ by increasing the number of active carbide sites for aromatization to take place.

At 700 C all of the three catalysts deactivated with time. This can be attributed to the generation of coke on catalytic surface. This was later confirmed by Temperature Programmed Oxidation (TPO) on the spent catalysts (not shown).

References:

(1) Spivey, J. J.; Hutchings, G. Chem Soc Rev 2014, 43, 792.

(2) F. J. Maldonado-Hódar, *Appl. Catal., A*, 2011, 408, 156–162

(3) Levy, R. B.; Boudart, M. Science 1973, 181, 547.

(4) Tshabalala, T. E.; Coville, N. J.; Anderson, J. A.; Scurrell, M. S. Applied Catalysis A: General 2015, 503, 218.

528000 Engineering Fe-HZSM-5 for Methane Dehydroaromatization

Abstract:

The abundance of domestic natural gas reserves has resulted in increasing interests in generating higher value chemicals from methane, the main component of natural gas. While methane upgrading is currently achieved through an indirect route via a syngas intermediate, direct upgrading offers conceptually a less energy intensive and costly alternative. One promising direct upgrading route is the non-oxidative conversion of methane to aromatics via dehydroaromatization (DHA). The reaction is typically conducted over catalysts composed of metal nanoparticles dispersed in zeolites, and zeolite channel size, the concentration of Brønsted acid sites, types of metal, and preparation methods are all known to influence the activity and selectivity for methane DHA. However, at the very high temperatures required for this reaction to proceed, carbon deposition becomes inevitable, results in rapid deactivation of the catalyst, and thus renders the process uneconomic.

Previous work in our group shows that atomically dispersed Fe could suppress coke formation on the metal cites while increase product selectivity toward benzene. In order to further mitigate coking on the zeolite acid site, here we investigated the effect of diffusion within ZSM-5 channel on minimizing undesired secondary reaction by tuning zeolite particle size and mesoporosity. We find that zeolite particle size can be controlled via the template/silica ratio, which also gives rise to a change in solution pH due to the alkalinity of tetrapropylammonium hydroxide solution (TPAOH) used as structure directing agent. The increase in alkalinity leads to higher mesoporosity. Balancing the pH by adding sodium hydroxide in turn affects the amount of sodium present during synthesis, which is also found to influence the mesopore volume of the sample. Careful and independent control of alkalinity, sodium amount, and TPAOH-to-silica ratio is hence required to isolate the effect of particle size and/or mesoporosity on activity and selectivity of the catalyst. Increasing mesoporosity is found to result in a strong increase in benzene selectivity and yield, while a change in particle size alone does not have a comparable impact on reactivity. This study hence demonstrated the effects of particle size and mesoporosity on the catalytic performance of Fe-ZSM5 in methane DHA and the ability to tune those parameters by carefully controlling template to silica ratio, alkalinity, and sodium amount, providing further insight into catalyst design.

526762 Magnesium Vanadate Catalyzed Oxidative Dehydrogenation of Ethane to Ethylene Using CO_2Â As a Soft Oxidant

Abstract:

In the United States, ethylene is produced by non-catalytic steam cracking of shale gas at temperatures as high as 800°C to yield ethylene, propylene, hydrogen, water, and other by-products. An alternative route for ethylene production is low temperature, highly selective, catalytic oxidative dehydrogenation (ODH) of ethane to ethylene in the presence of oxygen. The primary cost drivers for conventional ODH chemistry are the prices of inert gases and the use of pure oxygen. Our research has been focused on improving the process by lowering or eliminating the inert requirement and replacing oxygen with cheaper and safer CO_2. In this research, we examined the catalytic activity of various phases of supported magnesium vanadate catalysts for the CO_2 mediated ODH reaction. In addition, the effect of adding various promoters such as Mo, Co, Cr, and Fe were analyzed. Our studies indicate that magnesium pyrovanadate ($Mg_2V_2O_7$) is more active phase of the catalysts compared to magnesium metavanadate (MgV_2O_6) and orthovanadate ($Mg_3V_2O_8$) phase, which provides 12% ethane conversion, 48% ethylene selectivity, and 13% CO_2 conversion at 550°C. Chromium (Cr) has no effect on ethane conversion, whereas molybdenum (Mo) increase the selectivity. The combined effect of Mg, V, Cr and Mo provides 8% ethane conversion, 65% ethylene selectivity and 5% CO_2 conversion at 550°C without using any inert gas. The presentation will include detailed catalyst structure-activity correlations, reaction mechanisms, the effects of adding promoters and process economics.

536270 Chemical Looping for the Oxidative Cracking of Shale Condensates

Abstract:

Production of olefins by thermal cracking of light paraffins and naphtha is an energy intensive process (20-40 GJ_{th}/tonne ethylene). The burning of fuel required to meet this energy demand for olefin production makes cracking one of the single largest industrial contributors to anthropomorphic CO_2 emissions. In the case of naphtha cracking, 1.8-2 kg of CO_2 can be emitted per kg ethylene. To address the environmental challenges for olefin production, we have previously introduced several redox systems for the Chemical Looping Oxidative Dehydrogenation (CL-ODH) of ethane to ethylene. CL-ODH can greatly reduce the primary energy demand and CO_2 emissions for ethylene production, and it is well positioned to take advantage of ethane produced by the North American shale gas boom, cracking of heavier components is still an important concern. Not only is naphtha cracking the primary source of olefins in large parts of the world, ethane cracking does not produce the same amounts of industrially important C_{3+} olefins and di-olefin. Additionally, while shale-gas has increased the supply of ethane dramatically, shale-oil and gas have also flooded the market with condensates/ultra-light crude fractions that are unsuitable for use as transportation fuel. To take advantage of this market shift while simultaneously decreasing the energy demand and environmental impact of naphtha cracking, we propose a two-step Redox (or chemical looping) Oxidative Cracking (ROC) scheme where hydrogen, produced from the cracking of light liquid paraffins, is selectively oxidized by lattice oxygen from a redox catalyst. Regeneration of the lattice oxygen in a subsequent step heats the redox catalyst, providing the thermal energy needed for the cracking reaction. This process leads to minimal parasitic energy loss compared to cryogenic air separation and significantly reduced CO_2/NO_x emissions, while promoting olefin formation through the removal of hydrogen. In the current study, a promising mixed-metal-oxide based oxygen carriers are demonstrated for ROC. The general reaction pathways are elucidated. Simulations are also presented that indicate that the ROC process can reduce the overall energy demand and CO_2 emissions by half, while yielding increased C_{3+} products vs. ethane cracking or CL-ODH. Simulations also show that energy savings in the system significantly exceed the loss of hydrogen byproducts.

540146 The U.S. Department of Energy's Fossil Energy Research and Development Priorities

Abstract:

The Department of Energy's Fossil Energy Research and Development (R&D) program is geared toward meeting the Administration's goals of U.S. energy dominance, security, strong domestic energy production, and clean coal technology development.

Coal remains a critical energy source. However, the nation's aging coal-fired power plants require upgrades to make them more efficient and, therefore, lower emitting; and to ensure that they can operate on a grid that is increasingly accommodating intermittent, renewable generation.

At the same time, to compete with other sources of power generation, future coal-fired power plants will need to be of modular design, highly efficient, flexible, reliable, less expensive to build, and strategically located to provide grid stability; they must also have short start-up times and high turndown.

The Department's coal R&D is prioritized to meet these challenges under the proposed Advanced Coal Energy Systems (ACE) and Carbon Capture, Utilization, and Storage (CCUS) program. This program is focused on improving the efficiency and reliability of the existing aging fleet of coal-fired power plants, while developing advanced technologies and processes that will be used in the next generation of coal power plants.

Additionally, the Department continues early-stage R&D on CCUS, as well as research into new value-added products from coal—including rare earth elements from coal and coal by-products.

The phenomenal growth in U.S. oil and gas development is changing the energy landscape domestically and globally. To ensure the continued strong production and development of these resources, the Department's oil and gas R&D is focused on addressing the challenges associated with the production of resources from low-permeability formations, as well as midstream challenges, such as pipeline infrastructure and product delivery efficiency and flexibility. The Department has several pathways to meet these challenges, including technology development and the expanded use of data analytics, machine learning, and high-performance computing.

540293 Low-Carbon Solution to Xtl&c and Its Progress of Key Technologies

Abstract:

As the coal-rich country, coal or other carbon resources to liquid fuel and chemicals (XTL&C) sounds to be highly predominant in China, in particular their production at the scenario of low-carbon foot-print. Thus, low-carbon solution via hybrid energy system is being investigated from both geographical and sector view, and then key technologies R&D are carried out in terms of green carbon science. As a result, CO_2/CO utilization, as the central part of C1 chemistry, becomes one of the most severe challenges for human society nowadays. During the utilization of carbon-based resources, CO is the important intermediate molecules, but CO_2 is the end-product indicating the use-up of carbon energy. Among them, syngas to fuel and chemicals via FTS has attracted the attention from fundamental research to industrial deployment.

In chemistry, CO_2 utilization plays a role to complete the redox cycle, which is intrinsically redox reactions in nature. Charge transfer is the key step for the occurrence of the reactions. Moreover, charge transfer capability is critical for developing the related catalytic materials. As for CO utilization, the selective control becomes the key issue for syngas conversion into fuels or chemicals. Herein, nano/sub-nano catalysis plays a key role in the development of new technologies for CO_2 and CO utilization, and four key factors play the key role for selective catalysis, i.e. size, surface, surrounding and scale. Reaction network during syngas conversion concerning CO_2 and CO utilization demonstrate the importance of C-O bond activation by nano/sub-nano catalysis with respect to the factors above mentioned.

Based on those fundamental research concerning the nature of active site, reaction network and structure-performance relationship, both catalysts and process are being developed for XTL&C with non-ASF product distribution from tunable syngas composition via multiple methane reforming.

540112 The Global Energy Challenge: What's Ahead

Abstract:

The world is confronted by a two faceted energy challenge: on the one hand global energy demand is projected to grow significantly by mid-century and beyond, driven primarily by population growth and by economic growth in developing countries. At the same time, meeting the threat of climate change requires decarbonization of the entire energy system. In this presentation, I focus primarily on the power sector to illustrate addressing this dual energy/climate challenge.

Two examples of major changes in generation in the power sector – shale gas and solar energy – are used to illustrate what has been done to date in growing and decarbonizing electricity system. Each of these comes with its own challenges that together help define a long term strategy captured at MIT in a set of eight low-carbon energy centers. I give examples of game changing energy technologies emerging in these areas as illustrations of what needs to be done. Finally, developing countries present a unique set of challenges associated with additional economic and other constraints. The Tata Center for Technology and Design at MITEI provides a model for what can be done in these critical regions of the world.

535978 Advanced Modeling and Optimization for Future Generation Energy Systems

Abstract:

Optimization models are essential tools for decision-making. They are encountered in all facets of process engineering, from model and process development, to process synthesis and design, and for process operations, control, scheduling and planning. Unfortunately, engineering practice views most optimization strategies as search techniques that are wrapped around a simulation model. This glorified case study approach can be very expensive to execute, and it is often not clear whether excellent solutions are overlooked.

A more efficient approach is to *directly solve* the optimality conditions as an extension of the process model. This nonlinear programming (NLP) approach ensures that the best solutions have been found and that firm knowledge of neighboring solutions can be assessed. Moreover, sensitivity of the optimal solution to exogenous inputs, and robustness of the solution to uncertainties can be assessed with negligible cost through sensitivity analysis. Finally, an efficient and reliable optimization strategy integrates well within the work process and easily adapts to model variations and extensions. These features are addressed by leading edge process optimization modeling environments. Embedded within state-of-art optimization modeling tools, these approaches provide optimal solutions to real-world problems at far less cost (often orders of magnitude) than simulation-based approaches. In particular, a recent optimization framework has been developed by the Institute for the Design of Advanced Energy Systems (IDAES) at NETL, using Pyomo, a Python-based optimization modeling environment. The resulting platform leads to a powerful decision-making capability for large, complex energy systems.

Armed with these capabilities, this talk discusses recent advances in the integration of optimization into the engineering workplace. First, equation-oriented optimization models couple state of the art NLP solvers to parallelizable structures, and realize fast and reliable convergence behavior, in contrast to simulation-based optimization approaches. This has led to the solution of problems with potentially millions of variables and thousands of degrees of freedom. Second, enabling NLP tools handle optimization models with nonsmooth switches as well as phase changes in equilibrium systems. This capability allows the modeling of phase changes and complex phenomena in distillation columns, pipelines, complex heat exchangers and reservoir models. Finally, detailed (and expensive) procedures for device-scale (e.g., CFD) and physical property models derived from computational chemistry (DFT, molecular dynamics) resist the integration within system-wide optimization models. Instead, reduced models (RMs) are often substituted for decision-making tasks, but often with compromises on the optimality of these decisions. Recently, RM-based trust region frameworks have been developed that guarantee convergence to the optimum of the original detailed model (ODM), through the solution of RM-based trust region strategies with recourse to ODM evaluations. All of these features will be illustrated by challenging optimization case studies presented in the talk.

529863 High Purity Hydrogen Production in a 10 kW_{th} Fixed Bed RESC Prototype System

Abstract:

Abstract

The reformer sponge iron cycle (RESC) provides a suitable concept to realize high-purity pressurized hydrogen from a broad range of renewable resources as e.g. biogas or gasified biomass from wood or residues in decentralized on-site applications [1-3]. The on-site production of renewable hydrogen allows the realization of an infrastructure for low emission fuel cell based power production and mobility. The reformer steam iron process is introduced as the technological option to realize on-site hydrogen production by combining production, purification and storage with a low environmental impact [4,5].

A prototype system was developed, constructed and operated to produce hydrogen from syngas with a purity of 99.999% and enabling hydrogen output of up to 10 kW_{LHV} [2]. The design includes a high-temperature reactor with a combined steam reformer and a chemical looping section for hydrogen purification, comprising of up to 18 kg iron-oxide based oxygen carrier material. A series of experiments was conducted in with an enhanced process layout and analytic systems to improve the feasible hydrogen purity and determine the influence of several process parameters on hydrogen quality. In the chemical looping section, significant influences of the exothermic oxidation and endothermic reduction reaction are illustrated with axial and radial temperature profiles. Highly sensitive gas detection facilitated the detection of carbonaceous impurities in single digit ppm range. Furthermore, process analyses were performed to further improve the conversion efficiency according to thermodynamic simulations and earlier experiments in a smaller prototype system.

The results exhibit a very high hydrogen purity with total carbonaceous impurities (carbon monoxide, carbon dioxide, methane) below 10 ppm and a hydrogen output of up to 10kW_{LHV}. The conversion stability of the applied oxygen carrier material was reviewed in both single-particle thermogravimetric experiments and a special fixed-bed thermogravimetric device. Sintering effects and mechanical degradation of the pelletized material are illustrated and analyzed with SEM, EDX, XRD, light microscopy and BET methods.

Acknowledgement

The authors gratefully acknowledge financial support by the Austrian Federal Ministry of Transport, Innovation and Technology (BMVIT), Austrian Federal Ministry of Education, Science and Research (BMBFW) and the Austrian Research Promotion Agency (FFG) through the energy research program.

[1] S. Nestl, G. Voitic, R. Zacharias, S. Bock, and V. Hacker, "Operation of the reformer steam iron process prototype for hydrogen production," J. Energy Technol., 02 March 2018.

[2] S. Bock, R. Zacharias and V. Hacker, "High purity hydrogen from 10 kW_{th} RESC-prototype system", 2018, in preparation.

[3] G. Voitic, S. Nestl, K. Malli, V. Hacker, et al., "High purity pressurised hydrogen production from syngas by the steam-iron process," RSC Adv., 2016, 6, 58, 53533–53541.

[4] G. Voitic, V. Hacker, "Recent advancements in chemical looping water splitting for the production of hydrogen", RSC Advances, No.6, 10 October 2016, S. 98267-98296.

[5] V. Hacker, "A novel process for stationary hydrogen production: the reformer sponge iron cycle (RESC)", J. Power Sources, Vol. 118, Issues 1–2, 25 May 2003, Pages 311-314.

532079 High Purity Syngas and Hydrogen Coproduction from Natural Gas Using Cu-Fe Based Metal Oxides in a Chemical Looping System

Abstract:

The increased availability of natural gas as a cheap and readily available resource in the future makes it an ideal raw material for the petrochemical industry. Production of syngas and hydrogen is important because the former is a critical building block for valuable commodity chemicals and the latter is a source of clean energy in addition to being a widely used reactant. Chemical Looping technology presents an attractive alternative to the conventional processes for chemicals and energy production from natural gas with significant reduction in carbon emissions and high exergy efficiency. In a chemical looping system, metal oxide particles undergo sequential reactions, while circulating through different reactors to produce the desired products. In this paper, a three reactor chemical looping system is proposed where syngas is produced from methane in the reducer reactor, the reduced metal oxide particles from the reducer are then partially oxidized by steam to produce hydrogen in the oxidizer reactor and finally they are fully oxidized via air in a combustor reactor. Cu-Fe based oxides were investigated for use in the above mentioned process where different compositions of CuO and Fe_2O_3, with Al_2O_3 as the support, were tested to find a balance between their high reactivity towards natural gas and high selectivity towards syngas, respectively. The reactivity and recyclability of all the compositions was first tested in a Thermogravimetric analyzer (TGA) and then their selectivity towards syngas was tested in a U-tube fixed bed reactor. The Cu-Fe composition producing the highest syngas yields was demonstrated to produce high purity (97.5% dry basis) syngas and high purity (>99.5% dry basis) hydrogen while undergoing multiple reduction-oxidation (redox) cycles in a U-tube fixed bed reactor. X-ray diffraction (XRD) and Brunauerâ€“Emmettâ€“Teller (BET) surface area and pore volume analysis was done to determine the chemical and morphological changes in the metal oxides at different stages of reaction in the redox cycle. The advantage of using a co-current moving bed reactor for syngas production using Cu-Fe oxides was also demonstrated through experiments in a simulated fixed bed reactor. Experimental results from the fixed bed experiments were used in ASPEN Plus simulation software to perform detailed mass and heat balance and compare the overall energy efficiency of the chemical looping system against the conventional auto-thermal reforming (ATR) process.

535425 Process System Analysis of a High-Pressure Chemical Looping Based Hydrogen Production System

Abstract:

The iron-based moving bed based chemical looping technology developed at the Ohio State University (OSU) is a promising candidate to reduce costs associated with Steam Methane Reforming (SMR) for H_2 production. The OSU system performs the natural gas utilization and H_2 production in two separate reactors. As a result, it eliminates the need for multiple processing units to purify H_2. Initial estimates have shown an improvement in hydrogen production (2.4 moles of H_2 per mole of carbon in natural gas for chemical looping as opposed to 2.2 moles of H_2 per mole of carbon in natural gas for SMR system)[1]. The earlier chemical looping investigations quantified the thermodynamics of the chemical looping system for atmospheric pressure. So, while the earlier results are promising, it should be noted that various H_2 utilization applications require H_2 at pressures of ~30-50 bar.

This study seeks to quantify the thermodynamic feasibility associated with high-pressure chemical looping applications for Hydrogen production. This study initially quantifies the H_2 production potential of the chemical looping system for higher pressures under isothermal operating conditions. Identified optimal isothermal operating conditions are used for simulating a commercially relevant adiabatic reactor operation. Investigation of variation in the pre-heats of natural gas, the supports (wt%) to synchronize the endothermic reducer reactor and the exothermic combustor reaction show maxima in trends associated with each reducer inlet temperature and system operating pressure. The study will investigate the individual and synergistic effects of variables like the number of compressor and expander stages, variation in reactant space hourly velocities and the trade-offs associated with using pinch and transshipment type technology for heat exchanger network synthesis. Preliminary results show that operating chemical looping system at higher pressures is feasible. An autothermal production achieved at pressures of 32 and 64 bar while maintaining a better performance than the conventional SMR systems and atmospheric chemical looping systems.

1. Kathe, M. V., Empfield, A., Na, J., Blair, E., & Fan, L. S. (2016). Hydrogen production from natural gas using an iron-based chemical looping technology: Thermodynamic simulations and process system analysis. Applied Energy, 165, 183-201.

528304 Methane to Syngas By Chemical Looping Using FE-NI Oxygen Carriers: Reactor Design and Process Modeling

Abstract:

Chemical looping reaction scheme involves separation of a reaction into two separate steps, either spatially or temporally. One of the reactants, usually oxygen, is circled between the two steps using a metal-based oxygen carrier. Applications of chemical looping include total or partial oxidation of fuels. Typical oxygen carriers use transition metals such as Fe, Cu and Ni. Partial oxidation of methane (POM) to syngas (mixture of CO and H_2) using chemical looping has shown several advantages over the conventional process such as steam methane reforming (SMR) or catalytic partial oxidation (POX). The Fe-based oxygen carriers have the advantage of high selectivity to CO and H_2, while having low reactivity. On the other hand, Ni-based oxygen carriers show high selectivity towards total oxidation (production of CO_2), but have high reactivity. In previous work done by our group, novel oxygen carriers ($Ni_xFe_{(1-x)}$-CeO_2) were developed which combined the high selectivity of Fe for partial oxidation and high reactivity of Ni to produce syngas by methane reduction. $Ni_{0.12}Fe_{0.88}$-CeO_2 was identified as the optimal choice that maximizes the production of syngas while minimizing the formation of CO_2 and coking.

In this paper, we focus on a systems-level analysis of a methane-to-syngas process utilizing a chemical-looping fixed bed reactor with mixed Fe-Ni oxygen carrier material mentioned above. The overall system is designed to produce a desired flow rate and composition of syngas (as measured by the amount of CO and H_2 and the ratio of CO/H_2) from natural gas, while maximizing the yield and minimizing the overall energy consumption and CO_2 emissions. The system involves heating of natural gas and air (for reduction and oxidation steps, respectively), chemical looping reactor with alternating redox cycles, downstream cooling of gases with heat recovery, steam generation from waste heat and optionally, using steam for electricity generation. The model involves a detailed chemical looping reactor model, incorporated into the systems-level model. Reactor-level mass and energy balance calculations will be performed to calculate the effect of operating conditions such as flow rates, solids conversion and temperature on the final product composition, selectivity and yields of different components and solids inventory. The systems-level model includes thermodynamic calculations of individual process components such as blowers, compressors and heat recovery equipment, as well as steam turbines. The overall system efficiency and CO_2 emissions will be calculated for a design syngas flow rate and composition. Using sensitivity analysis on the process parameters, we will identify operating conditions required to improve process performance. The results will be compared to the conventional alternatives such as SMR and POX processes.

539325 Dry Reforming of Methane over Ni Based Lanthanum Zirconate Pyrochlore Catalysts: Deactivation Study

Abstract:

Dry reforming of methane (DRM) is one promising method for conversion of methane and carbon dioxide to syngas ($CO+H_2$). Typically, nickel-based catalysts are widely used in methane reforming due to availability and cost but are prone to deactivation by coking and sintering[1]. Various types of catalysts have been studied to optimize conversion and minimize deactivation. Pyrochlores are thermally stable crystalline metal oxides with general formula $A_2B_2O_7$ and are tolerant at methane reforming conditions. In our work, a modified Pechini[2] method was used to prepare Lanthanum Zirconate pyrochlore catalyst with Ni doped in Zr sites to produce 3wt%Ni catalyst. Spectroscopic and thermal characterizations of the catalyst were carried out.

Due to low Ni loading, XRD spectra did not show significant difference to the base pyrochlore spectra, except for few small peaks of NiO. In XPS, Ni^{+2} and Ni^{+3} were detected from Ni3p and O1s peaks as reported by Voevodin et al.[3] TPR of fresh catalysts showed a large peak above 550°C (Fig shown), which can be attributed to highly dispersed and strongly interacting Ni species with the support. Two other small peaks were also observed at lower temperatures which could be due to possible reduction of Ni^{+3} to Ni^{+2} and Ni^{+2} to Ni^0[4]. A temperature programmed reaction was carried out to observe activity and possible deactivation of the catalyst. Catalyst was reduced under hydrogen at high temperatures to activate the catalyst prior to the reaction. Around 700°C, the reactants, CO_2 and CH_4 started to react. Temperature was held for 90 min each at 700°C, 750°C, 800°C and 850°C and then brought back to 700°C. During this process, the product stream was monitored continuously using a Mass Spectrometer. Initial experiment at 700°C showed 93% and 88% conversions of CO_2 and CH_4 respectively. These conversions are higher than the those obtained for 4%Ni/Al_2O_3 from MónicaGarcía-Diéguez et al.[5], and 5%Ni/CeO_2-ZrO2 from Kambolis et al.[6], at the same operating temperature of 700°C. As the temperature was increased, increase in both CO_2 and CH_4 conversion values was observed and reached equilibrium at around 850°C. When the reaction temperature was brought back to 700°C, there was a significant drop in the conversion values indicating catalyst deactivation. Post reaction TPO showed two peaks indicating carbon deposition (Fig shown). Peak at 345°C can be assigned to amorphous carbon, while the peak at 610°C can be assigned to polymeric carbon. SEM images also confirmed carbon depoisition.

526321 Pressure Dilution, a New Method to Prepare a Stable Ni/Fumed Silica Catalyst for the Dry Reforming of Methane to Produce Hydrogen

Abstract:

Pressure dilution, a new method to prepare a stable Ni/fumed silica catalyst for the dry reforming of methane to produce hydrogen

V. Danghyan, S. Calderon Novoa, A. Mukasyan and E. E. Wolf

Department of Chemical and Biomolecular Engineering, University of Notre Dame, Notre Dame, IN

One of the main limitations of using the dry reforming of methane (DRM) on Ni catalysts is its deactivation due to the carbon formation [1]. For SiO_2 supported Ni catalysts, due to the weak metal-support interaction, Ni crystallites have high degree of reduction that results in high initial activity, but also in favorable conditions for carbon formation, eventually resulting in catalyst deactivation [2]. The main objective of this work was to find the pathways between C formation, catalyst structure, and catalyst deactivation with the aim to develop catalysts with increased activity and stability. In this context, a novel method of catalyst preparation, using pressure dilution, has developed, which increased the metal dispersion and stability of Ni/fumed SiO_2 catalysts.

Catalysts prepared by impregnation of $Ni(NO_3)_2{\cdot}6H_2O$ onto fumed SiO_2, were subjected to various pretreatments which yielded high initial activities at 600°C. The activation energy was measured to be 91 kJ/mol and the initial rate was of 8.5 [mol of CH_4/ g_{Ni} h]. Depending on the pretreatment, however, the catalysts deactivated at different rates due to carbon formation. The carbon structure was studied by SEM/TEM; is was found that carbon accumulates mostly in the form of carbon nanotubes (C-NT) with Ni crystallites at the top, where the reaction takes place without affecting the activity during first hours of TOS. There is also formation of clumps of entangled C-NT with Ni crystallites encapsulated by the entangled C-NT, leading to loss of active area and deactivation.

A newly discovered method of catalyst preparation, referred as pressure dilution (fig. 1), resulted in a significant increase in Ni dispersion from 19% to 61%, when supported on fumed SiO_2. Moreover, it was found that Ni dispersion can be controlled by the pressure applied during catalysts' preparation.

Fig. 1 Illustration of catalyst preparation methods and their effect on the metal dispersion and carbon formation during DRM.

The higher dispersion resulted in a higher catalyst activity and increased stability towards carbon formation. A hypothesis of the re-dispersion process occurring during p-dilution is proposed.

References:

(1) M.C.J. Bradford, M.A Vannice, Catal. Rev. 41:1 (1999), 1-42.

(2) J.W. Han, J.S. Park, M.S. Choi, H. Lee, Appl. Catal. B 203 (2017) 625-632.

533188 A Novel Carbon-Resistant Perovskite Catalyst for Hydrogen Production Using Methane Dry Reforming

Abstract:

The aim of this study is to evaluate a perovskite catalyst, prepared by the sol-gel method, as a carbon resistant Catalyst in methane dry reforming under different conditions. The ABO3+Î´ (CeCoxNi1-xO3+Î´) perovskite was used for the methane dry reforming investigation due to their crystalline nature and the ability to accommodate a wide range of cations, perovskites possess catalytically advantageous oxidation and reduction properties. In order to develop perovskites with potential for industrial application, high surface area oxides with composition homogeneity were selected. The Perovskites as multicomponent oxides was prepared by sol-gel synthesis which is considered as a most effective technique for preparing these compounds. Several characterization techniques were employed such as XRD,TEM,SEM,TPR,TPO and TOC to evaluate the properties of the fresh and spent perovskite catalyst. The tested perovskite catalyst ABO3+Î´ , oxygen-rich catalyst, exhibited superior catalytic performance over previously studied catalyst ABO3 such as LaCoO3 and LaNiO3 systems. It is believed that the high catalyst performance was achieved du to the electronic interactions between the Co and Ni sites which is a result of atomic scale level mixing of the ingredients using the sol-gel method. XRD, BET and TOC results revealed that the presence of ceria in the perovskite system used in methane dry reforming catalysts can improve coking resilience and confer stability even under prolonged usage in carbon-rich environments.

521261 Water-Gas Shift Reactor for Fuel Cell Systems: Stable Operation for 5000 Hours

Abstract:

Water-gas shift reactor for fuel cell systems: Stable operation for 5000 hours

Joachim Pasel*[1], Remzi Can Samsun[1], Andreas Tschauder[1], Ralf Peters[1], Detlef Stolten[1,2]

[1]Forschungszentrum Jülich GmbH, Institute of Energy and Climate Research, IEK-3: Electrochemical Process Engineering, 52425 Jülich, Germany

[2]Chair for Fuel Cells, RWTH Aachen University, Germany

The scientific work in the fuel processing and systems group at Forschungszentrum Juelich has the strategic aim of developing high-temperature polymer electrolyte fuel cell (HT-PEFC) systems based on autothermal reforming of diesel fuel and kerosene in the power class of 5 kW_e to 10 kW_e. One important component of the system besides the HT-PEFC stack and the autothermal reformer is the water-gas shift reactor (WGS). Its function is to significantly reduce the concentration of CO in the product gas flow leaving the autothermal reformer from approximately 7-10 vol% to the range of 1.0-1.5 vol%. CO poisons the anode reaction in the HT-PEFC working between 160 °C and 180 °C due to preferential adsorption and blocking of the active sites.

In this contribution the development of reactors for the water-gas shift reaction conducted in recent years at Jülich is described with a focus on technical advances in actual reactor generation denoted as 'WGS 6'. In addition, experimental results from a long-term experiment for 5000 hours on stream with WGS 6 are given and discussed with respect to the temporal trends of product gas composition and reactor temperatures within this timespan. For WGS 6, a specific layout was developed to increase the values for the power density and specific power. The layout of the WGS 6 is shown in Figure 1. The reformate emanating from the upstream autothermal reformer flows into the high-temperature shift (HTS) monolith, whose catalytically-active precious metal coating promotes the reduction of the CO concentration in the reformate. According to the design, the temperature level is ideally approximately 400-450 °C. After leaving the HTS monolith, the hot reformate molecules hit the underside of a curved deflection surface (hemisphere), onto whose topside a pressure swirl nozzle directs a fine spray of cold (20 °C) water droplets. Via the deflecting surface, the enthalpy flow of the hot reformate molecules is transferred to the cold water droplets hitting the topside of the hemisphere. Thereby, one minor part of the water droplets directly evaporates when it comes into contact with the hot surface. The majority of the water droplets, however, form a film on the deflecting surface. This film is then moved into the annular gap between the HTS and low-temperature shift (LTS) stages. Steam produced during initial contact with the hot deflecting surface then serves as carrier gas. At the end of the annular gap, the reformate flow is again deflected by 180 ° and enters the monolith of the LTS stage at a temperature of approximately 300 °C. Following the LTS stage, the reformate leaves the WGS 6 with a temperature of about 330 °C and can be fed into the anode of the HT-PEFC. The improved layout of the WGS 6 leads to a strongly increased power density of 12.2 kW l^{-1} and a specific power of 11.2 kW kg^{-1}. Thereby, WGS 6 can significantly contribute to decreasing the weight and volume of the fuel processing unit of an HT-PEFC fuel cell system.

Figure 1 Basic layout of the Jülich water-gas shift reactor WGS 6

Figure 2 displays the dry concentrations of H_2, CO_2 and CO as a function of time on stream during the long-term experiment. Under the reaction conditions defined by the outlet gas composition of the autothermal reformer and its temperature (340 °C), the dry concentrations of H_2, CO_2 and CO at the thermodynamic equilibrium amount to 40.7, 19.8 and 1.0 vol%, respectively. It is obvious that the concentrations of H_2 are fairly stable during the entire long-term experiment and amount to approximately 40 vol%. The concentrations of CO_2 are approximately 19.7 vol% at the beginning of the experiment and then slightly and continuously decrease to values close to 19.0 vol%. This difference of 0.7 vol% can be identified in the concentration trend for CO in a reversed way according to the stoichiometrics of the water-gas shift reaction. The concentration of CO amounts to 0.8 vol% at the start of the long-term run and increases to approximately 1.5 vol% by the end. It is noteworthy, that the concentrations of H_2, CO_2 and CO at the beginning of the experiment are very close to their values at thermodynamic equilibrium. During the entire long-term experiment with WGS 6, CO concentrations are within or below the advisable range of 1.0-1.5 vol%.

Figure 2 Dry concentrations of H_2, CO_2 and CO as a function of time on stream

530714 Enhancements on Diesel Autothermal Reformer for Low-Temperature Solid Oxide Fuel Cell Integration

Abstract:

A Diesel autothermal reformer has been improved to integrate with low-temperature Solid Oxide Fuel Cell (SOFC) stack for Auxiliary Power Unit (APU) applications. The Diesel reformer can provide hydrogen-rich reformate to the SOFC stack with a hydrocarbon reforming reaction. For the low-temperature SOFC stack, considerations on sulfur removal and heat management have to be done with the design process of the Diesel reformer. Liquid hydrocarbon fuels such as Diesel commonly have a small amount of sulfur content, and it has to be removed for stable operation of the fuel cell stack. In particular, for the low-temperature SOFC stack, the sulfur contents have to be reduced to less than ppm level. The desulfurizer can be located after the Diesel reformer to reduce the sulfur level, but its operating temperature is around 350-400°C, which is less than the operating temperature of the SOFC stack. Therefore, the reformate has to be re-heated to be supplied to the SOFC anode to ensure the stable operation. Heat exchangers can be designed to utilize waste heat from the SOFC stack and provide the heat to the reformate. For the low-temperature SOFC stack, the burden of the heat exchangers can be decreased, and overall system size can be reduced. With those considerations, enhancements on Diesel autothermal reformer for the low-temperature SOFC have been studied in order to improve overall system efficiency. A prototype of the 1 kW_e Diesel reformer system has been developed and tested to validate the improvements for the low-temperature SOFC application. In result, 950 W of electricity could be produced with the system.

524080 Rapid Production of High-Purity Hydrogen Fuel from Liquid Hydrocarbon Fossil Fuels

Abstract:

Hydrogen as an energy carrier promises a sustainable energy revolution. However, one of the greatest challenges for any future hydrogen economy is the necessity for large scale hydrogen production not involving concurrent CO_2 production. The intrinsically high hydrogen content of fossil fuels offers a potential route to CO_2-free hydrogen production through their catalytic deep dehydrogenation.

We have recently discovered a new approach to rapidly liberate high-purity hydrogen by *microwave-initiated catalytic dehydrogenation* of liquid alkanes using Fe and Ni particles supported on silicon carbide. A H_2 production selectivity from all evolved gases of some 98%, is achieved with less than a fraction of a percent of adventitious CO and CO_2. Such a rapid and selective production of hydrogen from liquid hydrocarbon alkanes reveals a powerful new platform for hydrogen production.

Importantly, these findings represent a new and attractive path towards a decarbonised fossil fuels economy: Thus, fluid hydrocarbon fuels are not combusted - but rather utilised - as an abundant, widespread source for the production of high purity hydrogen for use in fuel cells. In this approach, CO_2 would never enter the atmosphere as in the conventional combustion process. The utilisation process yields elemental hydrogen and elemental carbon; the latter can then be readily subjected to recycling through a range of established chemical routes.

530057 Hydrogen Production from Liquid Hydrocarbons with Process Intensification â€“ a Case Study

Abstract:

Driven by the need to conserve fuel resources while minimizing the carbon footprint of energy consumption, higher efficiency and lower emission technologies are being developed for both transportation and power generation markets. Among the various technology options being investigated, hydrogen emerges to become an important component of the future energy mix. One of the key challenges for hydrogen to reach an acceptable level of market penetration is to achieve significant reduction in the production cost. One approach is to reduce the capital cost by reducing the complexity of the process and thus minimizes the equipment needed to generate hydrogen.

A process concept that integrates all key operating (reaction, separation, and purification) steps involved in the hydrogen production process in a single reactor is being investigated. The overall integrated hydrogen production process using liquid hydrocarbons involves a pre-reforming step to produce a methane-rich stream followed by steam reforming and separation using a catalytic membrane reactor. Selective removal of hydrogen from the reaction environment using membranes also allows overcoming the thermodynamic equilibrium limitation for the reversible hydrogen production reactions. The membrane reformer then reforms the mixture and produces high-purity hydrogen via membrane permeation. The thermodynamic limitation is overcome by continuous removal of hydrogen from the reaction environment. The integrated process design will be compact due to elimination of water-gas shift reactors and pressure swing adsorption with improved process efficiency and reduced carbon footprint. In this study, iso-octane and heavy naphtha feeds were reformed in a lab-scale reactor to acquire preliminary insights on the industrial pre-reformer. The process optimization for operating parameters was also conducted on the entire system by building a process model in Aspen Plus. The lab-scale pre-reformer used a Ni-Ru/CGO catalyst and had a temperature range of 500 - 600°C, steam-to-carbon ratios of 2.5 and 3.0, gas-hour-space velocities (GHSVs) from 5,000 to 10,000 h^{-1}, and pressure from 1 to 30 bar. The process model and experimental findings will be discussed during the presentation.

539336 The Effect of Catalytic Metal on the Activity of Lanthanum Zirconate Pyrochlore Under Low S/C Methane Reforming Conditions

Abstract:

The effect of catalytic metal on the activity of Lanthanum Zirconate pyrochlore under low S/C methane reforming conditions

533088 Hydrogen Production through Diesel Steam Reforming: Process Intensification Using Structured Thermo-Neutral Reforming Catalyst

Abstract:

Hydrogen production is gaining increased importance from the viewpoints of ultimate clean and high caloric energy sources, and especially for the use in fuel cells. Hydrogen demand in the refining sector is also expected to continue its growth driven mainly by stricter fuel specifications. In addition, hydrogen-based fuel cells for automotive and stationary applications are gaining popularity for various reasons including their higher efficiencies and lower emissions. Use of liquid hydrocarbon fuels to generate hydrogen is being considered as an immediate solution for large scale hydrogen production. Diesel steam reforming is known for posing a major challenge due to its high carbon content, aromatics, and sulfur, which play a major role in catalyst deactivation. Similarly, the requirements for on-board hydrogen production demand a robust and highly efficient, compact, and cost effective reaction system for fuel processing. For this purpose, a multi-components thermo-neutral reforming (TNR) catalyst in various sizes and shapes was used to convert diesel into hydrogen rich syngas for fuel cell based power applications. Starting from powder form to pallets, monolith and finally microlith (brand name form PCI, USA) coated TNR catalyst was prepared, characterized and evaluated for syngas production. Monolith or microlith based structured catalysts are expected to withstand the rugged environment in which these fuel processers on-board vehicles are subjected for applications. In this study, a process intensification approach was adopted to achieve complete conversion of diesel to hydrogen rich syngas using microlith based TNR catalyst.

536492 High Temperature Catalyst Development and Testing for Low Cost and Efficient Solar Driven Sulfur-Based Hydrogen Production

Abstract:

To make the hydrogen economy realistic at a large scale, one of the main technical issues to overcome is relative to hydrogen production at low cost and without greenhouse emission. The Hybrid Sulfur (HyS) process, driven by solar power, has great potential to reach high efficiency and low hydrogen production costs without greenhouse emissions. The high temperature section of the HyS cycle, which operates the catalytic decomposition of sulfuric acid into sulfur dioxide, oxygen and water, is a fundamental section affecting the overall plant efficiency and capital investment.

Greenway Energy (GWE), in conjunction with the University of South Carolina (USC) and Idaho National Laboratory (INL), is developing and testing new catalytic materials to decompose sulfuric acid. A novel catalyst preparation technique, developed by USC, uses a combination of: (1) strong electrostatic adsorption (SEA), which permits formation of very small metal particles with a narrow distribution of sizes, and (2) electroless deposition (ED) to produce controlled bimetallic catalysts. This will result in limiting the catalyst deactivation by using very small particles of a high surface free energy core metal with a catalytically-active outer metal shell deposited by ED. This presentation will focus on the testing of these novel catalysts in the high-temperature sulfuric acid decomposition environment.

531886 Technoeconomic Study of Advanced H_2 Production Technologies: Membrane-Supported H_2O Splitting, Thermochemical Redox H_2O Splitting and Fuel-Assisted H_2O Electrolysis

Abstract:

In this presentation, we discuss three advanced hydrogen production technologies, (i.e., membrane-supported water splitting, thermochemical redox (chemical looping) water splitting and fuel-assisted water electrolysis) and compare them with the conventional steam methane reforming (SMR) process for high purity hydrogen (>99.95% purity) production. Membrane-supported water splitting utilizes a mixed ionic-electronic conductive (MIEC) membrane, (e.g., $La_{0.9}Ca_{0.1}O_{3-}$î· and $BaFe_{0.9}Zr_{0.1}O_{3-}$î·) to separate the water splitting products (i.e., O_2 and H_2), so that thermodynamic equilibrium of water splitting is shifted and higher hydrogen production rates have been achieved in the literature [1] [2]. Thermochemical redox water splitting uses an oxygen carrier (OC, e.g., CeO_2 [3]) to transfer the oxygen species. In the OC oxidation reaction, water is converted into lattice oxygen and hydrogen gas, while in the OC reduction reaction, the lattice oxygen is released or consumed. For the fuel-assisted water electrolysis process, hydrocarbons, (e.g., CH_4) is added on the anode side of the electrolysis cell to drive the water splitting process. Less electricity will be consumed compared with the conventional electrolysis cell and higher first law efficiency can be achieved [4].

First, we compare the thermodynamic efficiency of these technologies with renewable heat source at 282°C (540°F) based on the system-scaled thermodynamic models. The membrane and redox systems have similar first law efficiency compared with the SMR system, all reaching around 88%. Yet the electrolysis system has lower efficiency of ~83%, and electricity consumption accounts for around 31% of the total energy consumed. Secondly, the renewable heat accounts for 28.2%, 36.4% and 34.7% of the total energy consumption for membrane, redox and electrolysis systems, respectively. But the conventional SMR system could hardly utilize the renewable heat source as the overall system is exothermic. In addition, we compared the energy costs (i.e., fuel and electricity costs) of these systems (the cost for the renewable heat source is neglected). Results show that both membrane and redox systems have lower energy costs than the SMR system, decreasing the costs by 9% and 14%, respectively. However, electrolysis has 24% higher energy costs than the SMR system, due to the high price of the electricity (using averaged US electricity costs as the base case). We also investigate the CO_2 emissions and found that all the advanced systems emit less CO_2 than the conventional SMR system. The redox system has the largest CO_2 emission abatement potential with 12% emission decrease compared with the SMR system. The membrane and electrolysis systems also decrease CO_2 emissions by 7% and 5%, respectively. Furthermore, the total as-spent cost for these three advanced systems for a small-scaled hydrogen plant of 100 kmol H_2/h will be estimated to compare their costs of high purity hydrogen production.

Reference

[1] Wu, X.-Y., Ghoniem, A. F., and Uddi, M., 2016, "Enhancing co-production of H2 and syngas via water splitting and POM on surface-modified oxygen permeable membranes," AIChE J., 62(12), pp. 4427-4435.

[2] Dimitrakopoulos, G., Schucker, R. C., Derrickson, K., Johnson, J. R., KopeÄ‡, K. K., Shao, L., Alahmadi, F., and Ghoniem, A. F., 2017, "Hydrogen and Ethylene Production through Water-Splitting and Ethane Dehydrogenation Using $BaFe_{0.9}Zr_{0.1}O_{3-}$î· Mixed-Conductors," ECS Trans., 80(9), pp. 181-190.

[3] Zhao, Z. L., Uddi, M., Tsvetkov, N., Yildiz, B., and Ghoniem, A. F., 2016, "Redox Kinetics Study of Fuel Reduced Ceria for Chemical-Looping Water Splitting," Journal of Physical Chemistry C, 120(30), pp. 16271-16289.

[4] Luo, Y., Shi, Y., Li, W., Ni, M., and Cai, N., 2014, "Elementary reaction modeling and experimental characterization of solid oxide fuel-assisted steam electrolysis cells," Int. J. Hydrogen Energy, 39(20), pp. 10359-10373.

536198 Microwave Chemistry

Abstract:

The National Energy Technology Laboratory (NETL) is pioneering an effort to better understand and utilize the application of microwave (MW) fields to various chemical reactions and materials processes. The potential cost benefits and energy efficiency of MW-enhanced processes have yet to be fully explored and quantified. The benefits of MW application to reaction systems include rapid, selective heating of the catalyst surface and reacting species to promote conversion to desired products at significantly lower, bulk system temperatures. Although the benefits of microwaves are known, the nuances of how the material interacts with the EM field and the effect of variables such as frequency, power, and catalyst composition are not well understood. NETL has built a state-of-the-art REACT facility including MW characterization equipment and two microwave reactors, a 2 kW fixed frequency, fixed-bed flow reactor; and, a one-of-a-kind 500 W variable frequency reactor in order to expand the fundamental knowledge microwave based reactions. Chemical reaction research being conducted includes Fischer-Tropsch (FT) synthesis, methane dehydroaromatization, and coal pyrolysis. Catalyst development includes particle size, shape, and composition and characterizing the dielectric properties in order to enhance the interaction with the microwave field. This information, along with the MW field modeling in COMSOL that NETL is conducting will be used to design novel microwave-active catalyst materials and systems.

537784 Low-Temperature Microwave Plasma Conversion of Methane to Higher Hydrocarbons

Abstract:

Technological developments in the unconventional oil and gas industry incited explosive growth in production of shale gas and oil known as the Shale Revolution. The resulting availability of the low-cost unconventional gas in the United States has enabled competitive use of ethane -- a relatively minor fraction of natural gas -- as a feedstock for large-scale ethylene production. However, the bulk of natural gas – methane – continues to be used only for low-value heat and electricity generation and, to a significantly lesser extent, as a source of industrial hydrogen through steam-methane reforming. Converting methane into higher value chemicals could significantly impact the chemical industry by adding new supplies of raw materials. However, methane activation remains a major challenge.

H Quest has demonstrated feasibility of applying a microwave continuous flow reactor to high-rate conversion of methane to hydrogen, acetylene, and ethylene. In the experimental set-up, the process gas mixture passes through a resonant cavity reactor, where a microwave discharge is created through an energy input of up to 4kW. The discharge creates a plasma environment and ruptures the C–H bond in methane. Resulting methyl, hydrogen, and other radicals participate in the rapid self-catalytic reaction, promoting activation of other methane molecules, as well as serving as precursors to C2+ chemicals, solid carbons, and free hydrogen. A slipstream exhaust gas with a nitrogen trace was analyzed with a 2-channel Inficon MicroGC to determine product distributions and conversion rates.

The product distribution and conversion rates were shown to be highly dependent not only on the energy density and process gas composition, but also on specific reactor configuration and geometry, as well as the gas flow patterns. Results (input energy requirements, methane conversion rates, and product selectivity) from the experimental sweep through process conditions and reactor configurations with and without a downstream catalyst bed will be discussed. A downward trend in energy requirements as methane concentration increased in the process gas mixture indicates a direction for process optimization.

Unlike other thermal decomposition approaches, this process does not rely on conventional (contact, convective, or dielectric) heating or use of thermal plasmas. Rather, it employs a microwave resonant cavity to create localized and high-energy reaction zones in the gas as it passes through the reactor's active zone. Microwaves enable volumetric, non-contact energy transfer to the reactant flow, which is not achievable with radiative or conductive heating in furnaces, by accelerating free electrons in the partially ionized low-temperature plasma. Through electron-molecule collisions, these electrons both transfer the microwave energy to methane molecules and help overcome the high activation energy required by the rate limiting step of hydrocarbon (methane) pyrolysis – the endothermic cleavage of C-C and/or C-H bonds. This results in rapid, direct conversion of methane to chemically active species under atmospheric pressure and mild bulk temperatures: at least 500 lower than conventional decomposition methods.

Acknowledgements

H Quest Vanguard, Inc. is a privately held technology company, based in Pittsburgh, Pennsylvania, focused on the development and commercialization of novel hydrocarbon conversion technologies. This material is based on work supported by the Department of Energy, Office of Science under award DE-SC0017227.

516068 Catalyst Design for Microwave-Assisted Dry Reforming of Methane

Abstract:

The dry reforming of methane (DRM) produces H_2 and CO (syngas) from CO_2 and CH_4. In this regard, DRM enables "carbon neutral" syngas production because it consumes CO_2 as a reactant, whereas conventional methane steam reforming generates over nine tons of CO_2 for each ton of H_2. Using CO_2 to produce commodity chemicals, such as H_2 and CO, can generate revenue to offset carbon capture costs, reduce the carbon footprint of fossil-fuel powered processes, and allow sustainable use of our country's fossil fuel resources. Unfortunately, the high operating temperatures of DRM (>800°C) introduce challenges compared with the lower-temperature, carbon-positive, methane steam reforming.

Recently, microwave-assisted catalysis has been proposed as an enabling technology for high temperature chemical processes. Unlike traditional thermal heating, microwaves can rapidly heat catalysts to extremely high temperatures, which allows reactors to utilize excess renewable energy on an intermittent basis (load follow) to promote traditionally challenging, thermally-driven reactions. Moreover, microwaves directly heat the catalyst material, and not the entire reactor body, which reduces many of the traditional heat management challenges associated with high temperature DRM. Microwave absorption is a function of the electronic and magnetic properties of a material, and a properly designed catalyst can function as a both a microwave heater and a reactive surface for driving the desired reaction. However, microwave absorption is extremely sensitive to the catalyst's chemical state and electronic structure, and effective catalysts must maintain microwave activity across a wide range of temperatures and in both oxidative and reductive environments.

We have constructed a single-mode microwave reactor and investigated microwave assisted DRM using conductive metal oxides. In particular, the conductive cobalt based perovskites (ex: $La_{0.8}Sr_{0.2}CoO_3$) are attractive because they readily convert microwave energy into heat. Additionally, perovskites have good thermal stability at temperatures in excess of 1000°C, unlike conventional carbon-based microwave absorbers which must be kept below 900°C to avoid pyrolytic carbon sublimation. The conductivity of these oxides allows them to inductively heat in the microwave similar to metals or carbon-based absorbers, but with the advantage of maintaining conductivity even in oxidative environments. We have demonstrated the rapid heating of the microwave DRM reactor and were able to ramp from a ground state to reaction temperature (>700°C) in less than 3 minutes with consistent on-off cyclability. Using such a system, temperatures of 850°C and over 50% DRM conversion with an almost ideal 1:1 CO to H_2 ratio have been achieved using less than 100W of applied microwave power, demonstrating the suitability of microwave catalysts for rapid, efficient, and carbon neutral syngas production.

536101 Techno-Economic Analysis of Direct Non-Oxidative Conversion of Shale Gas Via Non-Thermal Microwave (MW) Plasma Catalysis

Abstract:

Commercially, shale gas (mainly methane) is typically converted to higher value added chemicals like aromatics and olefins using a multi-step process. Syngas is first produced using reforming or autothermal reforming process followed by syngas conversion processes like methanol-to-olefins/aromatics, oxidative coupling of methane, or Fischer-Tropsch process. This multi-step route has a low carbon utilization efficiency since substantial quantity of carbon gets removed from the process in the form of carbon monoxide/dioxide during the aromatics/olefins synthesis steps. Moreover, the syngas synthesis step is very capital- and energy-intensive so a synthesis route that can avoid or reduce the number of steps is highly desired.

While it may be desired to convert shale gas to aromatics/olefins directly, direct conversion of methane, the main component of shale gas, is challenging because it is a very stable molecule and it cannot be easily activated. Methane has a strong C-H bond with a first bond dissociation energy of 439.3 kJ/mol [1], which indicates that high amount of energy would be needed in order to break the bond. Moreover, the C-H bond of methane is stronger than higher hydrocarbon products and as a result these products are more reactive than methane [2]. Existing approaches for direct conversion of methane to higher hydrocarbons are thermal and catalytic pyrolysis, oxidative coupling and non-oxidative conversion of methane [3]. However, all these synthesis routes suffer from low methane conversion, high reaction temperatures (>800°C) and pressure, low product selectivity, high coke formation and catalyst deactivation [3]. Contrary to these conversion routes, non-thermal microwave plasma catalysis will enable direct non-oxidative conversion of shale gas at atmospheric pressure and under mild reactor conditions (>400°C) with high product yield. To the best of our knowledge, there are no studies in the open literature that has developed a model for this process. In this work, first a detailed first-principles model of the MW plasma reactor for aromatics/olefins production is developed. A plant-wide model is then developed to generate the products at desired specifications. Finally a techno-economic analysis of this novel process is undertaken. A number of sensitivity studies are done evaluating the impact of operational variables and investment parameters on key economic measures.

References

[1] Schwach P, Pan X, Bao X. Direct Conversion of Methane to Value-Added Chemicals over Heterogeneous Catalysts: Challenges and Prospects. Chem Rev 2017;117:8497â€"520.

[2] Holmen A. Direct conversion of methane to fuels and chemicals. Catal Today 2009;142:2â€"8.

[3] Alvarez-Galvan MC, Mota N, Ojeda S, Rojas S, Navarro RM, Fierro JLG. Direct methane conversion routes to chemicals and fuels. Catal Today 2011;171:15â€"23.

528933 Novel Microwave Assisted Ammonia Synthesis By Methane and Nitrogen Under Atmospheric Pressure

Abstract:

Microwave irradiation has been studied for a number of science fields such as chemistry and biology. Direct ammonia synthesis by using methane and nitrogen as feed is a novel process. However, activating the methane and nitrogen molecules is a big challenge. In this study, the methane-nitrogen ammonia synthesis was carried out under atmospheric pressure at a temperature of 600 °C. In addition, microwave irradiation was applied to this catalytic reaction system and more than 5000 ppm of ammonia was detected in outlet with co-production of hydrogen. The results show that the microwave irradiation not only is able to heat the catalyst effectively but also can activate the methane and nitrogen molecules. Also, microwave irradiation is an effective mean to generate plasma and it is observed that the existence of plasma enhances the yield of ammonia as well as the methane conversion. Thus, this study shows the potential of direct catalytic ammonia synthesis by methane and nitrogen under microwave irradiation.

529792 Spectroscopic and Microscopic Characterization of Nanographene Product By Microwave-Assisted Plasma-Mediated Methane Pyrolysis

Abstract:

Hydrogen is recognized as the most promising and environmentally benign energy carrier. In particular, it may be used as a transportation fuel with no pollution and with higher efficiency than the present fossil-based solutions. The current methods of producing hydrogen include steam methane reforming (SMR), coal gasification, water electrolysis, biomass gasification and thermochemical processes. Presently, 95% of industrial hydrogen is produced via steam-reforming of natural gas [1]. This process is burdened with high CO_2 emissions and water consumption [2], and successes in scaling it down to enable low-cost, distributed hydrogen production necessary for the hydrogen economy have been limited. The heavy environmental burden of this dominant process puts the two objectives of reducing carbon emissions and increasing the use of hydrogen for fuel in direct conflict.

Thermal decomposition of methane directly to carbon and hydrogen has been proposed as a viable, CO_2-free alternative process [3]. Carbon is 'naturally' captured as a solid product, and, provided its form has commercial non-combustive application, is sequestered and provides economic by-product credit for hydrogen co-production. Purely thermal decomposition of methane to hydrogen requires temperatures exceeding 1200 °C, which presents operational and energy source challenges, with the theoretical energy requirement of 78.4 kJ/mole CH_4 [4].

Use of catalysts has been shown to reduce the operating temperature requirements to as low as 700 °C [5]. However, heat generation requirement still heavily contributes to costs and to the environmental footprint, while efficient heat transfer to the gas continues to present a technical challenge. Furthermore, catalysts are quickly deactivated by the carbon deposits, requiring either catalyst regeneration (carbon burn-off), which further exacerbates the environmental footprint, or disposal, which harms commercial viability. Low-cost catalysts that don't require regeneration while retaining commercial value may be a solution, but haven't been shown at commercial scales.

H Quest's microwave plasma pyrolysis process presents a transformational solution to the challenge of efficient, clean, and cost-effective methane decomposition. Unlike other approaches, it does not rely on conventional (contact, convective, or dielectric) heating or use of thermal plasmas. Rather, it employs a microwave resonant cavity to create localized and high-energy reaction zones in the gas as it passes through the reactor's active zone. Microwaves enable volumetric, non-contact energy transfer to the reactant flow, which is not achievable with radiative or conductive heating in furnaces, by accelerating free electrons in the partially ionized low-temperature plasma. Through electron-molecule collisions, these electrons both transfer the microwave energy to methane molecules and help overcome the high activation energy required by the rate limiting step of hydrocarbon (methane) pyrolysis – the endothermic cleavage of C-C and/or C-H bonds. This results in rapid, direct conversion of methane to chemically active species under atmospheric pressure and mild bulk temperatures: at least 500 lower than conventional decomposition methods.

Additional microwave plasma process advantages include: rapid startup and shutdown, enabling use of intermittent renewable power sources; inherent modularity, scalability, and reduction of costs and risks through replication of a single high-throughput low-cost modular unit reactor; a higher level of safety thanks to lower temperatures and ambient pressures; and simplicity of power delivery [6]. With its prototype microwave plasma reactor, H Quest has demonstrated direct conversion of methane to hydrogen, higher-value chemicals (including acetylene and ethylene), and carbon products: conductive carbon black and a wide range of carbon/graphitic morphologies and nanostructures, including graphene.

Pristine graphene has superior properties of electrical, thermal properties, and mechanical strength. Realization of near-pristine graphene without oxides left over as byproducts of the graphene oxide form and only partially restored sp^2 aromaticity. In contrast to the well documented liquid-based processes, gas phase aerosol production requires no harsh chemicals or secondary processes to recover the sp^2 framework. Applications require material property characterization, particularly for new forms. Additionally, production scale up requires mapping process parameters to the product properties, in this case reactant H_2 and CH_4 concentrations to graphene physical-chemical characteristics.

HRTEM provides a visual map of product across an array of process conditions. Complementary characterization techniques provide quantification of product phase purity and quality. TGA differentiates carbon forms by oxidative reactivity – thereby assessing product purity and overall nanographene yield. TGA interpretation is aided by calibration with carbon standards possessing uniformity of morphology and nanostructure. XRD is applied to resolve carbon phases based on their different structure (lattice) parameters, differentiating graphene sheets versus amorphous carbon spherules and semi-graphitic particles. Effects of H_2/CH_4 ratio in the feedstream on quality and yield of nanographenes and other carbon forms will be discussed. The wide range of graphitic materials produced by our process in response to change in process conditions is a decisive improvement over the state-of-the-art methane decomposition processes.

Demonstrated production of a high-value carbon, nanographene from natural gas is the key, unprecedented breakthrough achieved by this microwave plasma process. While the graphene market is projected to continue to grow rapidly afterwards, approaching $2 billion by 2025, the extraordinarily high cost of conventional methods of production of graphene material is single greatest factor inhibiting the growth of the markets for these otherwise highly desired and versatile materials. With the 1-2 order of magnitude reduction in production costs we expect market growth to accelerate and new markets to open in industries where the use graphene has hitherto been cost-prohibitive. At present, structural applications are considered to be the largest growth area for this material, but given reports of e.g. five-fold increases in Li-ion battery capacity enabled by this material [7], energy storage may well be the next major market.

H Quest Vanguard is developing broad-spectrum microwave plasma processes targeting conversion of hydrocarbon feedstocks such as coal and natural gas to value-added materials, chemicals and fuels [8,9]. A microwave pyrolysis approach was originally proposed in response to the DARPA initiative for green-house gas (GHG)-emission-free and cost-effective production of US Air Force jet fuel from the domestic coal resources. In H Quest Vanguard's process, natural gas can be used in single-stage reactor as a hydrogen source, eliminating external hydrogen production units and the associated CO_2 production, water consumption, and capital costs, and providing excess hydrogen sufficient for downstream hydro-treating.

Acknowledgements

H Quest Vanguard, Inc. is a privately held technology company, based in Pittsburgh, Pennsylvania, focused on the development and commercialization of novel hydrocarbon conversion technologies. This material is based on work supported by the Department of Energy, Office of Science through sub-award agreement no. 186949 with H Quest Vanguard, Inc. under the Prime Award DE-SC0015895 Phase I STTR and through sub-award agreement no. 195532 with H Quest Vanguard, Inc. under the Prime Award DE-SC0017227 Phase I SBIR.

References

1. Ahmed, S., Aitani, A., Rahman, F., Al-Dawood, A., Al-Muhaish, F. (2009). Decomposition of hydrocarbons to hydrogen and carbon. *Applied Catalysis A: General*, *359*(1), 1-24.
2. Marbán, G., Valdés-Solís, T. (2007). Towards the hydrogen economy? *International Journal of Hydrogen Energy*, *32*(12), 1625-1637.
3. Muradov, N. (2000, May). Thermocatalytic CO_2-free production of hydrogen from hydrocarbon fuels. In *Proceedings of the 2000 Hydrogen Program Review, NREL/CP-570-28890. See also,* Muradov N. Thermocatalytic CO_2-free production of hydrogen from hydrocarbon fuels. U.S. DOE Hydrogen Program Review. U.S.: Department of Energy (DOE); 2002. NREL/CP-610-32405.
4. Abbas, H. F., Daud, W. W. (2010). Hydrogen production by methane decomposition: a review. *International Journal of Hydrogen Energy*, *35*(3), 1160-1190.
5. Amin, A. M., Croiset, E., Epling, W. (2011). Review of methane catalytic cracking for hydrogen production. *International Journal of Hydrogen Energy*, *36*(4), 2904-2935.
6. Dagle, R. A., Dagle, V., Bearden, M. D., Holladay, J. D., Krause, T. R., Ahmed, S. (2017). *An Overview of Natural Gas Conversion Technologies for Co-Production of Hydrogen and Value-Added Solid Carbon Products* (No. PNNL-26726; ANL-17/11). Pacific Northwest National Lab.(PNNL), Richland, WA (United States); Argonne National Lab, (ANL). Argonne, IL (United States).

537875 Performance of Microwave-Assisted Catalysis for Production of Upgraded Pyrolysis Bio-Oil from Poplar Species in a Continuous-Flow System

Abstract:

The aim of this study is to investigate the performance of a microwave-assisted catalytic process for upgrading pyrolysis oil. The bio-oil was obtained from ground poplar biomass in a continuous-flow rotative heating reactor, via pyrolysis at 500°C with a mean residence time of 15 minutes, as indicated by preliminary studies for maximum liquid production, under a 95%N_2/5%H_2 atmosphere. The exiting volatile stream was split three-way: one stream was immediately condensed in a cold-trap at substantive negative temperatures, the second stream was passed through a conventionally heated catalytic bed prior to flash-cooling and condensing, and the third stream was passed through a catalytic bed heated using microwaves. Different zeolite-based catalysts (HZSM5, Ni-ZSM5, Fe-ZSM5) were investigated to evaluate their performance under longer processing times. The catalytic bed temperatures ranged from (200 to 375°C), controlled via microwave power level and temperature controller, respectively. Upgraded bio-oil samples were collected at different time intervals (every 15 minutes) in order to evaluate the time-series performance of each catalyst for durations up to 3 hours. Bio-oil quantitative analysis was undertaken via a mass balance on the liquid, gas, and char, while bio-oil qualitative analysis was performed using GC-MS, CHNO, and bomb calorimetry. The results were compared with those obtained with conventional heating, and with the initial pyrolysis oil. Different catalyst â€“ to â€“ initial carbon ratios were investigated, and the catalysts were characterised with respect to surface properties, and surface activity, and extent of coking. Results indicate that microwave heating of the catalyst bed results in a better long-term performance of the catalyst with respect to its catalytic activity, with variations among the different catalysts used.

540034 Recovering Rare Earth Elements from Coal-Based Materials

Abstract:

In 2014, the Department of Energy, Office of Fossil Energy (DOE-FE) and the National Energy Technology Laboratory (NETL) performed an initial assessment and analysis of the feasibility of economically recovering rare earth elements (REEs) from coal and coal by-product streams, such as fly ash, coal refuse, and aqueous effluents. After reporting its findings in the DOE 2015 Report to Congress, the Department initiated a multi-year research and development program effort to demonstrate the technical feasibility and economic viability of separating and extracting REEs from domestic coal-based resource materials. Research efforts include design and construction of bench- and pilot-scale facilities to initially generate 2wt% REE pre-concentrates, and ultimately validation of 90-99%, high purity, salable, rare earth oxide (REO) production using conventional physical and chemical separation technologies in small pilot facilities by 2020. This presentation will provide an overview summary of DOE-NETLâ€™s REE Program and technical accomplishments over the past four years.

531391 Transmission Electron Microscopy Observations of Rare Earth Minerals in the Fire Clay Coal, Kentucky, and in the Derived Combustion Ashes

Abstract:

The Fire Clay coal, eastern Kentucky, is arguably the premier coal-based rare earth element resource in the eastern United States. While the high concentration is, in part, a function of the presence of a volcanic ash-fall deposit within the coal, the Dean (Fire Clay) coal in Knox County, Kentucky, does not contain the megascopically-visible ash-fall tonstein present in most other sections of the coal bed. Like the Fire Clay tonstein, a low-ash portion of the coal is enriched in rare earth elements (>2400 ppm, on ash basis). In addition to kaolinite produced in the diagenesis of volcanic glass, transmission electron microscopy studies indicate the coal contains primary kaolinite, La-Ce-Nd-Th monazite, barium niobate, native gold, and Fe-Ni-Cr spinels. The mineral assemblages, particularly the kaolinite-monazite association and its similarity to the tonsteins in the coal to the east, demonstrate the coal was subject to the REE-enriched volcanic ash fall, apparently just at a more dilute level than at locations where the tonstein is present.

Fly ash was collected from individual mechanical and electrostatic precipitator hoppers at a 130-MW pulverized-coal utility unit burning the Fire Clay coal. REE-bearing phases occur within fine minerals and, possibly, as part of the fly ash glass phase. Transmission electron microscopy was used to identify Y-Ce-Nd-La-bearing regions within a glassy matrix. Y- and U-bearing zircons were also found in the fly ash. Total REY occupies a narrow range from 719 to 775 ppm, showing an overall lack of partitioning of the REY elements with respect to particle size and temperature in ash collection systems. The heavy REE increase and the light REE/heavy REE ratio decreases from the mechanical hoppers through to the ESP rows.

532500 Selectivity and Recovery of Rare Earth Elements from Coal Ash Leachates Using Liquid Membrane-Based Separations

Abstract:

This research investigated the recovery of rare earth elements (REEs) from leachates of coal fly ash, a large volume waste material that is enriched in REEs. Here, we investigated two types of liquid membrane separation methods, liquid emulsion membrane (LEM) and a supported liquid membrane (SLM), for their potential to effectively concentrate REEs from fly ash leachates. This work focused on the selective of these processes for REEs relative to other major metal ions in leachates and the mechanism controlling REE selectivity. Both SLM and LEM processes demonstrated 3-5 orders of magnitude greater selectivity of individual REEs compared to major ions (e.g. Al, Si, Na). The LEM-based separation and conventional solvent extraction were most efficient for the light REEs (e.g. La, Ce, Pr, Nd, Sm, Eu, Gd) while the SLM approach was most efficient for the heavy REEs (e.g. Y, Dy, Ho, Er, Tm, Yb, Lu), suggesting different rate limiting steps for REE recovery. Future work will explore the alternative solvents and extractants for the process as a means to improve recovery efficiency of high value REEs.

535915 Effect of Calcium Halide Salt Addition on the Fate of Rare Earth Compounds during Coal Combustion Process

Abstract:

Calcium halides are sometimes added to coal prior to combustion to facilitate the removal of mercury from the flue gas. To assess the impact of these additives on the fate of the rare earth compounds (typically phosphates in the feed coal) through the power plant, it is convenient to study model systems of rare earth phosphate plus a calcium halide. In this work, the kinetics of the high-temperature roasting of lanthanum phosphate with calcium chloride and calcium bromide were studied at temperatures up to 1000° C. XRD was used to elucidate the reaction pathways observed both in the presence of air and in an oxygen-free environment. Apparent activation energy and Arrhenius pre-factor for each reaction were determined from TGA experiments.

535946 Examination of the Distribution and Form of the Rare Earth Elements in a Metalliferous Lignite Coal

Abstract:

According to the U.S. Department of Energy National Energy Technology Laboratory, rare earth elements (REE) provide significant value to our national security, energy independence, environmental future, and economic growth. REEs are utilized in a suite of high importance end-uses, such as cell phones, hybrid/electric vehicles, magnets, computer components, catalysts and many others. The unique magnetic properties of REEs, in particular, make them crucial materials for the growing renewable energy and electric automotive markets. REEs are used in the strongest permanent magnets currently known, and are used in generators for wind turbines and in electric motors. These same types of magnets, as well as other critical REE-based products are used in a host of military defense applications, as well.

China, in part due to its deposits of a unique REE resource (ion-adsorbed clays) that combines high quantities of the more valuable heavy and critical REE, as well as simple and low cost extraction, dominates the global supply market. In 2010, China established new quotas on exports of REEs, which resulted in huge increases in REE prices peaking in 2011 due to an expected supply shortage for critical applications. As a result, production at the California-based Mountain Pass Mine was re-started after several years of dormancy. However, after peaking in price in 2011, prices have dropped substantially to slightly above 2010 levels, challenging the profitability of non-China based production which consists mainly of hard rock deposits, that are deficient in critical REE and heavy REE. Some researchers have noted that mining of the Mountain Pass and similar resources will neither mitigate the crisis in REE resources nor eliminate the shortage of the most critical REE, but will only result in overproduction of excessive cerium.

China accounted for about 83% of the total global REE supply in 2016, down from about 95% prior to 2010. Meanwhile, the US production was zero, with the Mountain Pass mine having declared bankruptcy and closing operations in the last quarter of 2015. The U.S. is currently 100% import reliant for REEs. Although still dominating global supply of the heavy REE, some researchers have estimated that the Chinese ion-adsorbed clays resource will only last another 15-20 years. The Chinese clays represent essentially the entire global supply of heavy REE and most of the critical REE. Further, the bulk of Chinese reserves and production is from a hard rock deposit (Bayan Obo Mine) that contains only trace amounts of heavy REE, and supplies roughly 80% of the global light REE demand. Due to its limited supply, and because the Chinese clay resource is rich in heavy and critical REE, while most other traditional resources are deficient in these less common and more valuable elements, it is imperative that new domestic sources of REEs, especially the heavy and critical REE, be identified and processes be developed to produce them. Coal and coal byproducts have recently been identified as one of these potential new resources for REEs.

As part of the US Department of Energy effort to identify alternative domestic sources of Rare Earth Elements (REE), the University of North Dakota was awarded a project (DE-FE0027006) to determine the feasibility of recovery of REE from North Dakota lignite coal and related feedstocks. The project team includes the University of North Dakota,

Microbeam Technologies (MTI), Barr Engineering, Pacific Northwest National Laboratory (PNNL), and MLJ Consulting with the support of the cost share partners North American Coal Company, Great River Energy, Great Northern Properties, Minnkota Power Cooperative, the North Dakota University System, and the North Dakota Industrial Commission/Lignite Research Program. Technical support is also being provided by the North Dakota Geological Survey.

The project is currently in its final year of a 3-year effort and is focusing on demonstrating a novel technology for REE recovery from pre-combustion lignite coal and evaluating the overall process economics. Technical results to date have been highly promising, and the REE recovery process takes advantage of the unique properties of lignite in which the REE are weakly associated with the organic components in the coal. This is unlike higher-rank coals, such as Eastern U.S. bituminous coals that predominantly contain the REE associated in hard mineral forms. Locations in North Dakota have also been discovered that contain coals with REE concentration as high as 650 ppm (dry coal basis) or 2300 ppm (ash basis), which are exceptionally high levels.

Prior testing at UND has identified that the REE are enriched in the light specific gravity (SG) fractions of the lignite. Recently, an REE-rich lignite, containing about 260 ppm REE on a dry whole coal basis was subjected to multi-specific gravity float-sink separations. The goals of this effort were the following: **i)** SG fractions with highest REE concentration, **ii)** SG fractions with highest extractable REE content, **iii)** ash yield as a function of SG, **iv)** mineralogy as a function of ash yield, and v) mineral/REE liberation as a function of particle size and SG. The ultimate objective of this testing was to select the optimum SG range of the coal to use for the process development testing.

This presentation and paper will first provide an overview of UNDâ€™s efforts related to REE from lignite and the key results of technology development to date, and subsequently describe in detail the results of the above SG separations and analysis. The experimental testing and results described are not only important to the process development, but also provide new insights into the modes of occurrence of the REE and other important minerals/elements in the lignite. The results show that selection of the appropriate SG range results in a significant increase in REE concentration as well as an improvement in REE recovery and extraction selectivity. This type of approach can also be applied at the commercial-scale with conventional coal cleaning techniques.

527876 Economic Extraction, Recovery and Upgrading of Rare Earth Elements from Coal-Based Resources

Abstract:

Coal-based resources such as waste rock near the edges of coal seams often contain significant quantities of rare earth elements. These resources often end up on waste piles. In many locations, the waste piles become acidic after long periods of time, and the associated acidic environment causes a slow release of metals, including some rare earth elements that are lost in waste treatment facilities. These resources can be recovered economically through appropriate processing of the coal feed material to produce a waste that is concentrated in rare earth elements that can be extracted using heap leaching with solutions from engineered biooxidation of pyrite naturally found with the coal resource, and subsequently recovered through solvent extraction and precipitation. In order to address this opportunity for low cost REE extraction and recovery, the objective of this study and the related DOE-funded project is to demonstrate and improve methods to economically extract, recover, and upgrade the rare earth element (REE) contents from vast coal-based resources using integrated modeling, targeted coal preparation, biooxidation, solution conditioning, heap leaching, solvent extraction, and precipitation technologies to produce rare earth bearing products in an environmentally friendly way. Utilization of these integrated technologies will enable coal producers to utilize untapped resources to produce revenue and extend resource life while simultaneously reducing future environmental issues and costs as they supply critical domestic REEs. Although this approach will be designed for coal ore processing facilities, it will be easily adapted in the future to the recovery of REEs and mitigation of acid rock drainage from coal tailings. Thus, this technology will enable recovery of an underutilized resource in an environmentally friendly way. Initial results from this study will be presented and discussed.

520357 Hydrometallurgical Circuits for the Recovery of RARE Earth Elements from Coal Sources

Abstract:

Abstract

Rare earth elements (REEs) contained in coal and associated mineral matter exist in several forms including micro-dispersed minerals, ion adsorption onto clay surfaces, ion substituted in the inner layers of hydrated clays and chemically bound in minerals and the organic matter. An extensive test program found that the REEs can be effectively recovered from certain components of several coal sources by leaching under mild conditions followed by solvent extraction.

Individual core and channel samples collected from three coal seams were segregated into different sections based on their lithology. Each segment was evaluated for their REE content and further studied in batch leaching tests to assess the ability for REE recovery. The REE distribution was found to be enriched in certain segments such as specific partings and immediate floor material while other segments were low in REE content. The ability to recover the REEs by leaching also varied significantly which explained the results obtained on samples collected from process streams of active processing plants. REE recovery by leaching was found to be more favorable when treating mineral matter produced from the middle density fractions of a given coal as opposed to the coarse and fine reject process streams which were comprised of a composite of materials from roof, floor and partings.

A range of pH values were evaluated at a solution temperature of $25^{O}C$ and $75^{O}C$ using 1.2 M sulfuric acid. Nearly 85% recovery of the REEs was extracted from the middlings material collected from a plant treating coal from the Fire Clay seam. Acid leaching of a second middling source from a plant treating West Kentucky No. 13 material resulted in 60% of the total REEs being recovered under the same leaching conditions.

The leachates produced from coal sources contain relatively low concentrations of REEs in the range of 50 - 100 ppm while contaminant ions such as iron, aluminum, magnesium and calcium were present in concentrations measured in the thousands of ppm. The enrichment of the REEs was be achieved using either solvent extraction or selective precipitation. After either pathway, oxalic acid was added to the enriched REE solution which selectively precipitated the REEs from solution. The REE precipitates were roasted to produce a final RE oxide mixed concentrate.

For solvent extraction, the leachates were mixed into kerosene containing 1 M D2EHPA at an organic:aqueous ratio of 1:1 prior to a phase separation. A loading, scrubbing and stripping sequence resulted in a mixed concentrate containing around 50% REEs. Using the alternative approach involving selective precipitation provided a concentrate of 80% REEs or a 97% RE oxide mix. A pilot-scale plant is being construct to evaluate the processes and circuits when operated continuously.

520378 Determination and Recovery of Rare Earths from Coal Combustion By-Products

Abstract:

Everything in the earthâ€™s crust is present in coal, and combustion by-products, including coal ashes, have long been of interest for potential recovery of valuable metals and trace elements. The National Energy Technology Laboratoryâ€™s Research and Innovation Center recently initiated research for the recovery of rare earths from abundant domestic coal and coal by-products. The coal burning power plants are a domestic treasure chest for the trace elements, including economically important ones like the rare earths, and ash can be considered a low-grade ore for valuable metals. Most of the common inorganic lanthanide compounds, such as the phosphates found in coal, have very high melting, boiling, and thermal decomposition temperatures, allowing them to concentrate in combustion by-products such as bottom ash and fly ash.

Approximately seven hundred eighty six million tons of coal were mined in the United States in 2017, with most burned to generate electricity. With an average concentration of 62 ppm, it is estimated that 45,000 tons of lanthanide elements are present within the coals burned each year in US power plants. The partitioning of the rare earths across US coal power stations is being investigated through a collaborative research effort between the Electric Power Research Institute and the Department of Energy. Early results suggest the lanthanides mainly partition to the bottom ash and the fly ash. A brief overview of recent results for the determination and recovery of rare earths in various solid samples obtained across different points within coal-burning plants will be presented, and future research will be discussed.

520857 Evaluation of Critical Trace Elements Including Rare Earth Elements in U.S. Coals

Abstract:

Coal is a potential source of some valuable critical elements. In this work, concentrations of 40 critical elements in U.S. coals were evaluated using the COALQUAL Database Version 3.0 with the main objective to identify best coal sources for potential recovery of these critical elements. A method was proposed to calculate the mean concentrations of critical elements in U.S. coals. According to the calculated mean concentrations, a rough estimate indicates that U.S. coals contain a large amount of critical elements that are enough to meet U.S. demands for many years to come, if these elements can be commercially extracted. Results further indicate that despite some variations among elements, bituminous coals from the Appalachian region are likely to have high concentrations of Li, Ga, Be, Se, Sb, Y, and rare earth elements (REE), which make them a potential source of these critical elements. Bituminous coals from the Interior Coal Province, both Eastern and Western, were found to have relatively high probabilities of having high Ge concentrations. In addition, data quality will also be discussed.

535953 Treatment Approaches for Produced Water Re-Use and Surface Discharge

Abstract:

Treatment Approaches for Produced Water Re-use and Surface Discharge

Madison Wenzlick, Alexandra Hakala, Nicholas Siefert

Nearly all methods for extracting oil and gas resources produce a high salinity brine byproduct [1]. Half of the high salinity brines produced in the U.S. are currently managed by injection into salt water disposal wells, which is a practice that has been linked to increased seismicity and has prompted new regulatory policies. One option for reducing the amount of reinjected brine is to remove fresh water from the brine, which allows for development of a tailor-made brine of a desired density or TDS [2]. These near-saturation NaCl brines could be sold as a feedstock to the chemical manufacturing industry. For example, saturated NaCl brines are called "10-lb brines" and are a tradeable commodity in parts of the U.S, particularly in Texas and Louisiana. Dewatering produced brines from oil & gas wells creates both a concentrated, lower-volume brine and fresh water.

Figure 1: Relative locations of formations and basins, chemical plants, as well as locations of known salt dome collapses.

Base map from eia.gov

The main commercially available technology for concentrating brines is mechanical vapor recompression (MVR). However, typically the energy usage and cost of this technology are high and there is still limited data available on applying MVR to oil and gas produced brines.

In this presentation, the following topics will be covered. First, we will discuss the region (TX & LA) and shale formations chosen for this analysis. Second, we will present a geochemical analysis of the brines produced in the formations in this region. The analysis will include geographic considerations of brine concentration, determination of necessary pretreatment and chemical use, transportation, and potential benefits from reuse. Next, we will present a baseline techno-economic analysis of concentrating these brines into 10-lb brines while generating fresh water for surface discharge. The baseline case is a process using commercially-available treatment technologies, such as MVR.

Figure 2: The baseline process flow diagram for the centralized water treatment process

Next, we present results from a sensitivity analysis around a number of the assumptions in the model, such as the cost of pretreatment chemicals, electricity costs, and brine disposal fees. Finally, we present an advanced case in which we model advanced membrane processes for both the pretreatment and brine concentrating steps of the process. Preliminary results from the baseline case suggest that, for those brines with low concentrations of divalent cations (Ca^{+2}, Mg^{+2}, Sr^{+2}), there is a positive rate of return on investment for dewatering oil and gas produced brines in TX & LA into 10-lb brine and fresh water.

Figure 3: Advanced case process flow diagram

References

[1] J. Veil, "U.S. Produced Water Volumes and Management Practices in 2012," 2015.

[2] J. T. Arena, J. C. Jain, C. L. Lopano, J. A. Hakala, T. V. Bartholomew, M. S. Mauter and N. S. Siefert, "Management and dewatering of brines extracted from geologic carbon storage sites," *International Journal of Greenhouse Gas Control,* no. 63, pp. 194-214, 2017.

529250 Advanced Supercritical Water-Based Process Concepts for Treatment and Beneficial Reuse of Brine Generated By Oil/Gas Production

Abstract:

Both U.S. energy and economic security rely upon continued development of U.S. unconventional shale plays, which require access to suitably clean water resources. Produced water (brine) is the nation's largest industrial waste stream, with approximately 14 billion barrels generated annually by the U.S. oil and gas sector. The recent surge and continued development of unconventional U.S. shale plays will likely result in greater volumes of produced water. In addition, an additional brine waste stream will be created by future sequestration of CO_2 emissions generated by fossil-based power plants into saline aquifers. Utilizing this brine in development of unconventional shale resources represents a beneficial reuse of this waste stream. This approach can not only lower stress on local watersheds but can also address public/local government concerns regarding long-term ground water contamination potential and seismic activity associated with Class II salt water disposal wells (SWDs).

However, direct reuse of brines in shale development activities currently is limited due to constituents found in this waste stream. Specifically, the high levels of dissolved solids found in the brine may cause scaling within the shale or within production casings, reducing well productivity. Current brine treatment technologies including membrane- and thermal-based processes are ineffective in treating brine containing concentrations of dissolved solids greater than 80,000 ppm due to fouling or cost/sizing, respectively.

To address these limitations, Ohio University (OHIO) with funding from the U.S. Department of Energy's National Energy Technology Laboratory (NETL) Crosscutting Research Program Project DE-FE0026315, has been developing an advanced supercritical water (SCW)-based process for treatment and beneficial reuse of brine waste streams. This SCW process offers an advantageous media for brine treatment, as its lower fluid density and decreased hydrogen bond strength provides a means to simultaneously remove dissolved solids and hydrocarbons.

Previous SCW-based brine treatment systems have been plagued by internal scaling, resulting in inefficient heat transfer, plugging, and process downtime. To address this issue OHIO has been developing both externally- and internally-heated SCW reactor design concepts, which utilize advantageous fluid dynamics and electrically driven Joule-heating mechanisms, respectively. OHIO's new SCW reactor designs offer the potential to provide a field deployable brine treatment process and a resultant product which may be reused in shale development or other beneficial reuse applications, thereby supporting goals of NETL's Strategic Center for Oil and Natural Gas.

To evaluate process potential, OHIO has been conducting both experimental investigations using prototype SCW reactors and process simulations/techno-economic assessments using Aspen Plus^{TM}. Removal of naturally occurring radioactive materials (NORM) using commercial ion exchange resins and natural adsorbent has been investigated. Further, removal of dissolved solids from simulated and field-derived produced water has been investigated at pressures ranging from 23-28 MPa, respectively, demonstrating the ability to recover greater than 99 percent of salts from brines containing dissolved solid concentrations greater than 240,000 mg·L^{-1}. Techno-economic summaries have identified important brine treatment cost sensitivity parameters and heat recovery schemes to lower brine treatment costs. This presentation will review experimental and techno-economic study results from both U.S. DOE supported projects.

516908 Permian & Anadarko Basin Produced Water Recycling: Keys to Success

Abstract:

Demand for recycling of produced water in the Permian & Anadarko Basin are expanding. Although 2017 hasnâ€™t necessarily been a drought year, some areas remain short on available water. Moreover, the considerations that producers must account for is vastly expanded over prior years. Issues such as Induced Seismicity, limited disposal zones for Class II wells, water availability, costs for fresh water, environmental concerns, and more are all expanding interest in produced water recycling! However, there are a multitude of options available that can be used for recycling and many companies have spent years vast amounts of money trying to test or evaluate various technologies. This paper discusses a very basic and successful approach to water recycling in some key development areas of the mid-continent and associated keys to success. For instance, Class II Disposal Wells can play a critical role in making a recycling project successful.

526555 Catalytic Etherification of Glycerol to Glycerol Oligomers in the Presence of Alumina Supported Ca/Sr Mixed Oxides

Abstract:

In this work, the solvent-less etherification of glycerol to oligomers using alumina supported Ca and Sr catalysts was studied. These catalysts were obtained with impregnation of strontium and calcium onto γ-alumina (γ-Al_2O_3) in $Sr(NO_3)_2$ and $Ca(NO_3)_2$ solutions. The evidence for synergistic effects between active Ca and Sr species was demonstrated by etherification tests over a series of Ca-Sr-Al catalysts with varied Ca-to-Sr molar ratios. Effect of calcination temperatures of these supported catalysts, from 600 °C to 900 °C, on the resultant reusability and conversion efficiency of glycerol to oligomers and the selectivity of linear and branched oligomer to cyclic ones were investigated. The catalyst properties were characterized by various instruments including BET, TPD, XRD and SEM, and the ICP-OES analysis was performed to confirm the leaching of the metal species in the liquid phase of the reaction mixture. With Ca and Sr supported on γ-alumina catalyst, the catalyzed etherification reactions were carried out at 250 °C with 2.5 wt% Ca/Sr/γ-alumina calcined at 900 °C in nitrogen atmosphere. Under optimal impregnation conditions, a conversion efficiency of glycerol 80 % glycerol could be obtained after 10-h reaction. Moreover, a selectivity near 70% of di- and tri-glycerol in linear and branched forms over the cyclic form could be achieved.

538721 The Role of Copper Stability in Selectively Condensation of Ethanol to Higher Alcohols

Abstract:

The recent emergence of a strong bioethanol industry has elevated interest towards its use as a sustainable platform molecule towards producing high valued chemicals; one such process is the selective conversion of ethanol into high value C_{4+} alcohols. This multistep reaction involving acid-base catalyzed condensation and hydrogen transfer tends to yield a complex mixture of side products such ethers, esters, and ketones that prevent efficient conversion. Our work has demonstrated 80% selectivity to higher alcohols with conversion above on MgAl hydrotalcite catalysts at low copper loadings, with stable catalyst lifetimes over 100 hours in a flow through-reactor setup. Lower concentrations of copper was shown to prevent the formation of ester and ketone side products without reducing the conversion of ethanol. Characterization of the catalyst by STEM and XAS EXAFS work reveal that this is largely due to high dispersion of copper at lower copper loadings with near absence of copper clustering, even under high temperatures and reductive conditions. Further *in operando* XANES and EXAFS study shows that copper only reduces partially to Cu

support matrix allow for the production of dispersed Cu^{+1} sites, which promotes the efficient catalytic reaction to higher alcohols while not contributing to side reaction pathways such as esterification and C-C scission. The dispersion and stability of these individual copper promoters is critical to the high efficiency of this reaction. This work will discuss the limiting copper movement within the support to lengthen the lifetime of catalyst activity as well as throughout multiple regeneration cycles.

538786 Effects of Alloying Pd and Cu on Tandem Dehydrogenation-Aldol Condensation Reactions

Abstract:

The successful reduction of carbon dioxide emissions as a result of the production and consumption of fuels and chemicals requires the use of biomass resources as a feedstock. As a result of this, the discovery of novel pathways to upgrade platform molecules to fuel precursors is of critical importance. More specifically, for the production of diesel substitutes, the formation of new C-C bonds is required, as platform molecules deriving from the acid-catalyzed hydrolysis of biomass or the fermentation thereof are in the C2-C6 carbon range.

In this work, we investigate the upgrade of fermentation-derived acetone-butanol-ethanol mixtures to renewable diesel precursors. To achieve this, we develop a bimetallic Pd-Cu catalyst. We show that alloying of Pd and Cu results in the suppression of undesirable decarbonylation reactions. We present X-ray absorption spectroscopy results showing the segregation of Cu to the catalyst nanoparticle surface. Kinetic measurements under flow conditions demonstrate that Pd-Cu surface ensembles are active for hydrogenation reactions, while Pd-Pd ensembles catalyze C-C bond scission reactions. Density functional theory and microkinetic modeling show that the PdCu ensembles are bind CO fragments weakly, and hence destabilize C-C bond scission transition states. Similarly, the weaker CO binding onto low-index planes compared to stepped surfaces results in higher decarbonylation rates over smaller nanoparticles. Based on these results, we generalize our conclusions for a range of reactions, under oxidizing and reducing conditions, and present the implications of the segregation of alloy components to the catalyst surface on reactivity and selectivity.

References:

1. *Nature* 491 (2013) 235â€"239
2. *Am. Chem. Soc.* 138 (2016) 6805-6812
3. *Catal. Sci. Technol.* 8 (2018) 314-327

530475 Lubricant Base Oils Production from Biomass

Abstract:

The gradual depletion of world fossil fuel resources and global environmental concerns have prompted the use of biomass feedstock for renewable products. Lubricant is one such product that is primarily obtained from petroleum and it is introduced to reduce friction between surfaces in mutual contact. Base oils are the main components of lubricant and the majority of them are petroleum-derived mineral oils and synthetic oils (e.g. poly-Î±-olefins). The oils are formulated with different additives into several grades to suit automotive and industrial applications. Production of lubricant base oils from biomass will be an attractive alternative to petroleum derived-based oils to improve sustainability. So far, a few efforts have been devoted to the synthesis of lubricant-based oil alicyclic hydrocarbons, from biomass furans.[1-2] In this work, we report a novel route for selective production of base oil hydrocarbons from biomass platform chemicals and natural oils. The overall yield of hydrocarbons could reach up to 80%. Importantly, the hydrocarbon structures can be tailored using different biomass derived compounds to tune properties for various applications.

1. Balakrishnan, E. R. Sacia, S. Sreekumar, G. Gunbas, A. A. Gokhale, C. D. Scown, F. D. Toste and A. T. Bell, Proc. Natl. Acad. Sci. USA 2015, 112, 7645-7649.
2. Balakrishnan, G. E. Arab, O. B. Kunbargi, A. A. Gokhale, A. M. Grippo, F. D. Toste and A. T. Bell, Green Chem. 2016, 18, 3577-3581.

530529 Selectivity Control during the One-Pot Conversion of Aliphatic Carboxylic Acids to Linear Olefins through Tandem Hydrogenation/Dehydration

Abstract:

Olefins are important building blocks for the synthesis of higher value chemicals in the petrochemical industry and they are typically produced from fossil fuels. As an alternative to fossil fuel-based chemicals, biomass-derived carboxylic acids could serve as a viable replacement feed for olefin production. However, the production of olefins from carboxylic acids in one step remains a challenge due to the stability of carboxylic acids and the tendency of olefins to over-hydrogenate during hydrogenation, leading to alkanes.

In the present study, we have studied the one-pot catalytic conversion of linear aliphatic carboxylic acids to linear olefins via tandem hydrogenation/dehydration reactions. Hexanoic acid was converted to a mixture of hexenes over Cu nanoparticles supported on commercial silica-alumina (Siral 70) in a continuous flow reactor. At 483 K and 5 bar H_2 gauge pressure, we obtained 98.5% conversion of hexanoic acid, with a molar product distribution of 88.9% hexenes and 11.1% hexane. Remarkably, we observed a brusque selectivity switch to 99.9% hexane at 100% conversion.

A product intermediates study shows that when switching from a feed of pure hexene to a mixture of hexene and hexanoic acid, overhydrogenation of hexene can be suppressed. In addition, hexene temperature-programmed desorption (TPD) and FT-IR spectroscopy reveals that adsorption of hexanoic acid prevents hexene from accessing the catalyst surface, thus avoiding overhydrogenation. Reactions using butyric acid as feed also exhibit similar phenomena. We have also performed reactions utilizing butyric acid derived from biomass fermentation, albeit with significant catalyst deactivation.

We propose that the presence of a small amount of carboxylic acid on the surface of the catalyst prevents overhydrogenation of the olefins. In light of these results, current efforts are targeted at applying this concept to explain similar phenomena in other reactions, particularly in biomass conversion. The presence of numerous carboxyl groups in native/pre-treated biomass might explain the difference in catalyst performance/selectivity when processing model feed versus real biomass. Furthermore, the addition of carboxylic acids to the reaction medium during biomass upgrading could afford more control over the selectivity of the reaction.

526827 Mechanism and Kinetics of Isobutene Production over Zirconia-Supported Zinc Oxides

Abstract:

Growing interest in utilizing biomass-derived platform chemicals to produce fuels and specialty chemicals has motivated the investigation of the reaction of ethanol or acetone to form isobutene, a valuable specialty chemical used in the production of fuel additives, polymers, and other high-value products. Reports in the literature have identified zinc-zirconium mixed oxides as effective and selective catalysts for the production of isobutene from ethanol at 723 K. In this investigation, zinc-zirconium mixed oxides prepared via incipient wetness impregnation were selective and highly active catalysts for the acetone to isobutene reaction, reaching near-theoretical isobutene selectivities of over 86%. The mechanism and kinetics of the acetone and ethanol to isobutene reactions were studied in order to elucidate the reaction pathway, the roles of active acidic and basic sites, and the role of water in promoting stability and selectivity. Zinc-zirconium mixed oxide catalysts with varying zinc loadings were synthesized and characterized with XRD, BET, CO2-TPD, and DRIFTS FTIR with pyridine adsorption. Kinetic studies suggested that the ethanol to isobutene reaction proceeds via dehydrogenation to acetaldehyde, followed by condensation to form acetic acid, aldol condensation to produce acetone, dimerization of acetone to form diacetone alcohol, dehydration to form mesityl oxide, then finally hydrolysis of mesityl oxide to produce isobutene and acetic acid, which undergoes further aldol condensation to produce acetone. Understanding the reaction pathway, mechanism, and roles of the acidic and basic sites can help to define the catalyst properties required to selectively produce isobutene from a variety of sustainably-derived platform molecules.

533068 Catalytic Upgrading of Sugar-Derived Polyols to Glycols in Absence of Externally Added Hydrogen

Abstract:

Due to the increasing demand for fossil fuels and their diminishing supply worldwide, it is of considerable importance to utilize renewable resources for sustainable development of human society. Replacing fossil-based feedstock with biomass-derived ones to produce renewable fuels and chemicals is one of the promising ways to address challenges facing the energy and chemical industries. Sugar polyols, such as sorbitol, xylitol and erythritol are attractive platform chemicals because they can be readily derived from cellulosic biomass. Hydrogenolysis of sugar polyols can be converted to a variety of industrially important chemical intermediates including ethylene glycol (EG) and 1,2-propanediol (1,2-PDO), which are widely used as antifreezes, solvents and pharmaceuticals.

Up to date, it is known most existing hydrogenolysis processes are carried out under high temperature (T > 220 °C) and high H_2 pressure conditions (P > 4 MPa), which cause several technological issues such as poor product selectivity due to significant side reactions, fast catalyst deactivation as well as high capital cost and energy consumption.

In this work, we proposed a novel one pot H_2 generation and hydrogenolysis process in absence of external H_2 under mild conditions (T: 160 °C-200 °C, P: 1.0 Mpa nitrogen atmosphere).We have designed a series of bimetallic Pd-M (M: Cu, Ni, Pt, *etc*) catalysts supported on active carbon and layered double hydroxides and confirmed their remarkable performances in catalytic conversion of xylitol, sorbitol and mannitol to 1,2-PDO, EG and lactic acid (LA). In particular, catalyst activity and selectivity will be studied for different catalysts. Structure-performance correlations for H_2 generation and hydrogenolysis will be established based on bulk and surface characterization using XRD, SEM, TEM and chemisorption. The bimetallic Pd-Ni/AC and Pd-Pt/AC catalysts showed superior hydrogenated activities as well as selectivities towards lactic acid and glycols, compared with other bimetallic catalysts under identical conditions. At nearly 60.0% sorbitol conversion, a selectivity of 36.5% to ethylene glycol, propanediol and other glycols was obtained on Pd-Ni/AC at 200 °C, 1.0Mpa nitrogen atmosphere in absence of external H_2. Here we also researched reaction mechanism or pathways by studying metal active sites, H_2 spillover effect and the role of solid-base via regulating the growth of Pd-M and the ratio of acidity (basicity). It was concluded that H_2-spillover from Pd to M and solid-base acidity (basicity) may strongly influence the reaction mechanism to some extent, thereby influence the activity and selectivity of hydrogenolysis of biomass-derived to chemicals. Further details on reaction pathways and kinetic modeling will also be discussed in this presentation.

533584 Selective C-C Bond Scission of Primary Alcohols Using Cerium Oxide-Supported Palladium Catalyst

Abstract:

Biomass attracts significant attention as a carbon-neutral resource with the potential to suppress the CO_2 emissions. Recently, the biomass-derived chemicals such as glycerol, furfural, and levulinic acid are identified as the best suited alternatives for petroleum-derived chemicals. From the stand point of green sustainable chemistry, development of highly selective catalytic transformation of these compounds to valuable chemicals has been desired. To date, much effort has been devoted to selective C–O bond cleavage of high-oxygen containing biogenic polyols, such as glycerol, to produce valuable chemicals by hydrogenolysis and deoxydehydration.[1,2] On the other hand, C–C bond cleavage has not yet been widely researched despite its great potential to extend the utility of biomass-derived chemicals. The reported C–C bond cleavage methods often suffer from low selectivity toward the desired chemicals and limited substrate scope. In our group, we have reported the selective C–C bond scission of platform biomass-derived chemicals catalyzed by Ru/CeO_2 in the presence of water under 3 MPa of H_2.[3] In order to open new routes for industrially important chemicals from a wide range of biomass derivatives, development of highly selective and greener C–C bond scission catalysts is desired.

In this work, we found that cerium oxide-supported palladium (Pd/CeO_2) efficiently promoted the selective one-carbon scission of primary alcohols under air without H_2. The cleavage of C–C bonds occurs at positions adjacent to the hydroxymethyl groups (Scheme 1). The solid catalyst was reusable without appreciable loss of the activity and selectivity.

References

[1] T. Mizugaki, Y. Nagatsu, K. Togo, Z. Maeno, T. Mitsudome, K. Jitsukawa, K. Kaneda, *Green Chem.* **2015**, *17*, 5136–5139.

[2] T. Mizugaki, K. Togo, Z. Maeno, T. Mitsudome, K. Jitsukawa, K. Kaneda, *ACS Sustainable Chem. Eng.* **2016**, *4*, 682–685.

[3] T. Mizugaki, K. Togo, Z. Maeno, T. Mitsudome, K. Jitsukawa, K. Kaneda, *Sci. Rep.* **2017**, *7*, 14007.

533996 New Technologies to Enhance the Production of Unconventional Oil and Natural Gas

Abstract:

Title: New Technologies to Enhance the Production of Unconventional Oil and Natural Gas

Author: *Jared Ciferno (National Energy Technology Laboratory; jared.ciferno@netl.doe.gov)*

Abstract

Despite the significant growth in the production of Unconventional Oil and Gas (UOG) resources during the shale revolution, very little is known in detail about recovery efficiency in these hydraulically-fractured reservoirs. The limited information that is available suggests that recovery factors (the ratio of produced resource to total in-place resource) are typically quite low: approximately 20 percent in gas-rich shale reservoirs and less than 10 percent in liquid-rich plays. By increasing the recovery factors of hydrocarbons, the environmental impact of these activities would also be reduced.

The Department of Energy's National Energy Technology Laboratory is implementing an Advanced Unconventional Oil & Gas Technology Program to address key research challenges focused on increasing the recovery efficiency and reducing the environmental impact of unconventional oil & gas production. The programmatic research foci include: 1) implementing basin-specific field laboratories to evaluate novel methodologies for improving reservoir characterization, well completion, and enhancing hydrocarbon recovery; 2) developing technologies to better the nano-scale structure of UOG reservoirs in order to more effectively create and control stimulated rock volume; 3) developing technologies and processes that are capable of reducing water demand while improving recovery efficiency; and 4) accurately assessing the potential nature, scale, and rate (intensity) of future development in various areas of the country to guide assessment of region-specific and environmental impacts.

This discussion will provide an overview of these research elements and expected outcomes.

534253 Characterization of Microfractures in Organic-RICH Shales and Tight Reservoir Rocks of the Bakken Formation By Integrated Microscopy Techniques

Abstract:

An integrated microscopy approach has been employed to characterize microfracture properties of organic-rich shales and tight reservoir rock samples from the Bakken Formation in the Williston Basin, North Dakota. A combination of microscopic techniques, including field emission scanning electron microscopy (FESEM), optical microscopy thin-section analysis, and ultra violet fluorescence (UVF) microscopy, provides an opportunity to characterize hydraulic fractures in unconventional reservoir samples at the nano- and microscale. A determination of nanoscale properties, such as fracture aperture and fracture-fill minerals, can be extremely useful for optimizing hydraulic fracture operations in tight oil formations. Knowledge of fracture-fill minerals also assists in predicting geochemical interactions with hydraulic fracturing fluids or stimulation fluids that could lead to precipitation and, in turn, affect flow within the microfractures. The FESEM technique provides data on nanoscale fracture properties and chemical composition of fracture-fill minerals, while thin-section and UVF analyses allow property determination at the microscale; data from the two scales are interpreted jointly to better understand the impact on subsurface hydraulic fracturing activities.

Preliminary results obtained in this study indicate that the UVF technique has a lower microfracture detection limit around 1 Âµm, compared to a detection limit of about 2.5 Âµm for thin-section analysis. Also, about 40%â€“50% of the fractures detected by UVF are not detected via thin-section analysis. These detection differences might be attributed to the different light exposure modes: transmitted mode for thin-section analysis and reflected mode for UVF. Based on FESEM analysis, fracture-fill material was found to be precipitated salts, organic matter, and/or clays and pyrite. Detailed FESEM fracture morphology made it possible to distinguish between microfractures created in the subsurface during original coring and those created at the surface due to handling and sample preparation. Natural microfractures can also be determined based on their morphological attributes and fill material. These findings are useful for improving predictive reservoir modeling approaches and for enhanced understanding of optimal hydraulic fracturing and/or stimulation practices.

535561 Remediation of Polyacrylamide-Induced Permeability Damage from Fracturing Using H_2O_2 Catalyzed By Free and Immobilized Peroxidase Enzyme

Abstract:

Polyacrylamide (PAM) polymers are often employed in hydraulic fracturing fluids, especially in slickwater applications. However, they result in accumulation of "filter cake" on the surface of the formation which is known to lower oil or gas production efficiency. Existing chemical "breakers" may not fully degrade the filter cake or not work at the appropriate times or conditions. Here, we illustrate a novel and environmentally-benign approach for degrading partially hydrolyzed polyacrylamide (HPAM) with hydrogen peroxide catalyzed by the enzyme, horseradish peroxidase (HRP). The system was applied to both HPAM solutions and on HPAM filtercake on low permeability rock. This approach was found to significantly reduce the viscosity of HPAM solutions in pure water and brine. Molecular weight measurements indicated that the viscosity reduction was due to a significant degradation of the carbon-carbon backbone of the polymer and not due to amide hydrolysis or polymer rearrangement, etc. Preliminary results with HPAM-plugged nylon filters illustrate some flowrate recovery. Core flooding studies were conducted on low-permeability (~1 md) Indiana limestone as a model, and damage was partially remediated using free enzyme. The recovery was further improved using a novel approach by immobilizing the enzyme on a common fracturing proppant, Ottawa sand.

530664 Investigation of Scaling As a Means for Decreased Petroleum Production from the Utica/Point Pleasant Unconventional Play

Abstract:

The development of U.S. unconventional resources has had a dramatic impact on global energy supply, with the U.S. becoming the world's leading producer of petroleum and natural gas. This change has been beneficial to the public lowering energy prices, while increasing energy security. Unconventional (or shale) resources differ from conventional reservoirs in a multitude of ways. Production wise unconventional plays see significantly larger production drops, about 75% [1], after one year compared to 5% in conventional [2]. Unconventional wells also differ structurally and geologically as they consist of micro-pore and mainly nano-pore structures of interbedded carbonate and hydrocarbon containing shale layers which serve as the source for reactions with fracturing fluid. A majority of the reactions that occur during hydraulic fracturing materialize in the early stages, as the fracturing fluid has yet to come to equilibrium with the reservoir. While some reactions will create the pathways for hydrocarbons to flow through the well; others may cause the shale surface to be coated with insoluble precipitates. These insoluble solids tend to clog the fracture and pore networks creating large decreases in production. Prevention of scaling is vital for maintaining well integrity, which is usually achieved by adding scale inhibitors to the fracturing fluid. However in some wells, a scale inhibitor may be incompatible with well geology or ineffective once equilibrium has been reached resulting in scaling and lower production. Understanding, the mechanism of shale-fluid interactions will aid in identifying better hydraulic fracturing fluid chemistry to minimize or prevent formation of precipitates.

Ohio University (OHIO) is investigating scale formation as a means for the large decrease in production seen in unconventional wells, with a particular focus in the oil window of the Utica-Point Pleasant (UPP) Shale. In this work, shale-fluid interactions is investigated using experimental trials in a batch reactor at reservoir conditions. The effect of shale to fluid ratios to control reaction rates and scale inhibitors to prevent scale is also investigated. Pre- and post-trial analysis of the shale surfaces are performed using Scanning Electron Microscopy (SEM) paired with Energy-dispersive X-Ray spectroscopy (EDS) and Laser-Induced Breakdown Spectroscopy (LIBS), and fluids are analyzed using Inductively Coupled Plasma Optical Emission Spectrometry (ICP-OES) and Ion Chromatography (IC). Preliminary results in the absence of scale inhibitor indicate iron is a major concern for precipitation in the UPP oil window. These iron based scales are even formed at low pH values found in the hydraulic fracturing fluid chemistry. Additionally, Barium and strontium were found to be non-reactive. Furthermore, the effect of various commercially available scale inhibitors and different shale to fluid ratios will be presented. In addition, modeling of the shale-fluid interactions using Geochemist Workbench will be performed and presented.

[1] H. Wachtmeister, L. Lund, K. Aleklett, and M. Höök, "Production Decline Curves of Tight Oil Wells in Eagle Ford Shale," *Nat. Resour. Res.*, vol. 26, no. 3, pp. 365–377, Jul. 2017.

[2] M. Höök, R. Hirsch, and K. Aleklett, "Giant oil field decline rates and their influence on world oil production," *China Energy Effic.*, vol. 37, no. 6, pp. 2262–2272, Jun. 2009.

529292 CO_2-in-Mineral Oil Emulsions, CH_4-in-Mineral Oil Foams and N_2-in-Mineral Oil Foams Stabilized By Novel Oil-Soluble Surfactants As Waterless Hydraulic Fracturing Fluids

Abstract:

Objectives/ Scope: Viscous, completely water-free, high pressure liquid CO_2-in-oil (C/O) emulsions, methane-in-oil (M/O) foams and N_2-in-oil (N/O) foams have been stabilized by novel, interfacially active, oil-soluble, polymeric surfactants. For example, for (C/O) emulsions, the surfactant is composed of an oleophilic long-chain alkyl pendant groups on a CO_2-philic, oil-phobic, polydimethylsiloxane (PDMS) backbone. For (N/O) and (M/O) foams, neither N_2 nor CH_4 has appreciable solvent strength therefore the surfactant design is based upon oilphilic/oilphobic considerations. These emulsions and foams can serve as hydraulic fracturing fluids to eliminate the need for water and reduce formation damage caused by water.

Methods, Procedures: We designed and synthesized series of polymeric CO_2-philic PDMS- oil-phobic alkyl surfactants for (C/O) emulsions; polymeric oil-philic/oil-phobic surfactants were designed for the N/O and M/O foams. The emulsions/foams were formed by either mixing with an impeller (quality or volume fraction up to 60 vol%) or by simultaneously co-injecting the fluids through a packed bed of 22 Darcy unconsolidated sand (quality up to 90 vol%).

Results, Observations, Conclusions: The novel, completely hydrophobic surfactants adsorb at the (C/O), (N/O) or (M/O) interface despite the very low interfacial tension (driving force) without surfactant present. For C/O emulsions, at a concentration of 2 wt % in oil, the surfactant stabilizes the emulsions and no gelation of oil occurs. The 50-90% quality C/O emulsions, with CO_2 droplet sizes in the 5 â€" 150 micron range, exhibited an apparent viscosity of about 3 - 18 cP at 25°C and 2500 psia as determined with falling ball and capillary viscometers. The emulsions were stable for over a day as indicated by photographs and determination of the bubble size with microscopy. In contrast, N/O foams had an apparent viscosity ranging from 15-65 cP. Oil-soluble surfactants with optimum oilphilic/oil-phobic balance, based on the decrease in the interfacial tension, were the only surfactants that stabilized the N/O and M/O foams.

Novelty/Additive Information: We have addressed a major limitation in the generation and stabilization non-aqueous CO_2 and N_2 energized emulsions and foams for water-sensitive formations, by providing general criteria for surfactant design. The surfactant design criteria are based on a fundamental understanding of the relationship between foam apparent viscosity in terms of interfacial properties and phase behavior, as well as foam texture. We also compare C/O emulsions and N/O and M/O foams to understand the role of gas solubility in oil on apparent viscosity. The waterless, viscous emulsions and foams are hydraulic fracturing fluids that offer an efficient way to add both the novel surfactant and the proppant to the mineral oil at ambient pressure (rather than blending surfactant and/or proppant with high pressure CO_2 or N_2) for improved oil or gas recovery in water-sensitive formations.

524316 Hydraulic Fracture Propagation in Unconventional Reservoirs Under the Influence of Natural Fracture Heterogeneities

Abstract:

Determination of the hydraulic fracture propagation patterns in fractured unconventional reservoirs is important for the reservoir stimulation design and production estimation/enhancement. Past studies on hydraulic fracture-natural fracture interaction typically carry the assumptions that 1) natural fracture height coinciding with the reservoir/hydraulic fracture height, and 2) uniform properties of the natural fracture, which lead to 2D views. However, recent field studies and experimental observations demonstrate that natural fracture heterogeneities can strongly influence the hydraulic fracture propagation, which calls for further understandings. In this study, we consider the impact of natural fractures with varying mechanical and geometrical properties on hydraulic fracture propagation. A 3D analytical criterion is developed to estimate the dependence of hydraulic fracture propagation on spatially-varied natural fracture properties, characterized by the proportion of cemented region(s), cementation strength, and natural fracture height. It is compared with experimental results from analogue hydraulic fracturing tests and observed to provide good predictions. A lattice simulator is then employed to realize the fully-coupled simulation of hydraulic fracture growth in fractured media. Consistent with laboratory observations, the numerical simulation captures the 3D interaction behaviors between hydraulic fractures and natural fractures with non-uniform properties. It also matches the analytical criterion well for cases with different natural fracture cohesion, friction coefficient, tensile strength, location of cemented regions, cemented proportions, and/or host rock materials.

528653 Characterization ofÂ CO_2, Fluid, and Shale Via Feature Relocation Using Field-Emission Scanning Electron Microscopy, *in Situ* Infrared Spectroscopy, and Pore Size Analysis

Abstract:

Fundamental research examining the geochemical interactions of CO_2 and fluids with shale is limited from the perspective of understanding shale as a potential CO_2 storage reservoir and as an effective natural seal for hydrocarbons or CO_2. In addition, little is known about the effects of hydraulic fracturing agents such as water and CO_2 on the shale reservoir. Reaction of CO_2 with in-situ fluids, fracturing fluids, and reactive shale interfaces may generate new reactive surfaces or intermediates that may alter the properties of the formation. A fundamental understanding of the reactivity of CO_2 with shale and fluid interfaces will help in identifying 1) how CO_2 storage in shale formations plays a role in Carbon Capture and Storage (CCS) activities, 2) if CO_2 can be applied as a potential fracturing agent, and 3) whether CO_2 is an effective fluid for hydrocarbon extraction. In this work, we apply in-situ Fourier Transform infrared spectroscopy (FT-IR), feature relocation scanning electron microscopy (SEM), and Brunauer-Emmett-Teller (BET) surface area and density functional theory (DFT) pore size analysis to examine the effects of CO_2 and fluid on the shales across the U.S.

524736 Gas Phase Catalytic Oxidation of Lignin to Produce Phenolic Compounds over Vanadia Catalysts

Abstract:

The pretreatment of lignocellulosic biomass to create fermentable sugar streams *en route* to fuel/chemical production may also produce streams of insoluble lignin. These lignin stream can be burned to produce process heat and power. An alternative use of the lignin streams could be the production of valuable products such as phenolic compounds. This work explores the pyrolysis of lignin residue to produce gas phase products which can be passed of vanadia-based catalysts to oxidatively remove methoxy side chains to produce phenolic compounds, which could be used in the production of polymers and resins. Because lignin contains a complex mixture of linkages with methoxy groups, as well as both C-C and C-O bonds linking phenyl groups, there are number of bonds that must be broken in order to funnel lignin into the desired slate of phenolic compounds without methoxy groups (simple phenols). The subject of this presentation uses pyrolysis of lignin to create a mixture of gas phase species that can be passed over a catalyst to oxidatively cleave the methoxy groups to produce simple phenols. Because the goal is to partially oxidize the lignin pyrolysis vapors to produce phenols, as opposed to over-oxidation which would result in CO and CO_2 production, the oxidation activity of the catalysts must be controlled. Catalysts with varying vanadium-oxygen bond strength, which affects the oxidation activity, were designed by altering the catalytic support and synthesized using incipient wetness impregnation. Reaction experiments were conducted by passing lignin pyrolysis vapors over the catalysts at 500 °C and were shown to produce simple phenols from methoxy phenols. The effect of catalyst composition will be discussed along with relevant characterization techniques, coupled with computational modeling, to help understand the role of metal-oxygen bond strength in the catalytic activity.

526393 Mechanistic Study of the Hydrogenolysis of Diaryl Ethers Catalyzed By Heterogeneous Metal Catalysts

Abstract:

Manipulating reaction pathways of metal-catalyzed reactions is key to efficiently convert polar aromatic compounds. Diaryl ethers undergo C-O hydrogenolysis and hydrogenation with rates that depend on the specificity of the metal. We are studying the corresponding catalytic chemistry of Ni, Pd, Ru and Pt with reaction kinetics and experiments with isotopically labeled reactants. We have found a common reaction mechanisms for all studied metals but different isotopic effects, which points to the interaction of hydrogen with the metal as the factor determining selectivity. The details of the mechanisms, which are being tested by theoretical calculations, will be reported and selectivity rationalized as a function of the binding energy of reactants with the metal.

527899 Palladium-Iron Bimetallic Catalyst: High Activity and Stability for Aqueous Phase Hydrogenations

Abstract:

A novel palladium-iron bimetallic catalyst was synthesized that provided high hydrogenation activity and stability relative to palladium only under harsh hydrothermal conditions. To generate the catalyst, palladium was dispersed on carbon coated silica, which was doped with iron. The catalyst was applied to the hydrogenation of carbonyl groups in acetone, 2-pentanone, and propionaldehyde. The Pd-Fe catalyst gave more than a three-fold higher hydrogenation activity than the nanoparticle Pd only counterpart, while Fe nanoparticles were shown to be inert. The synthesis procedure was optimized for the bimetallic material, which demonstrated that the introduction of Fe contributed to a higher turnover frequency rather than simply increases the number of active sites. Interestingly, the same synthesis procedure was used to synthesize a Pd-Mo bimetallic catalyst that also increased activity.

536491 Hydrodeoxygenation of Guaiacol over Ni and Mo Nanoparticles Supported on SBA-15 and γ-Al_2O_3.

Abstract:

1. **Introduction**

The limitation of fossil energy resources and the increasing need for chemicals and liquid fuels require strong efforts in utilizing renewable energy sources. In this way, a technology such as biomass to liquid fuels (BTL) is being focus. The fast pyrolysis is a promising tool to transform various biomass feedstock's, lignocellulosic biomass, into an energy-dense and oily mixture. However, it possesses considerable amounts of oxygenates molecules, phenolic compounds and water in the material obtained by this method make it impossible to use the final product as motor fuel. For the use in chemical industry or fuel production an upgrading of the pyrolysis oil is necessary, which can be achieved via hydrodeoxygenation (HDO). This process consists in the treatment of bio-oil at temperatures between 200 and 450 °C with hydrogen at high pressure (10–300 bar) in the presence of a catalyst, so that the oxygen contained in organic molecules can be extracted with water formation[1]. Research efforts to study the chemistry of hydrodeoygenation of biomass-derived oil are gaining considerable importace over the last 20 years[1].The most frequently examined HDO catalysts commercially used for petroleum hydrotreatment products are $NiMo/Al_2O_3$ and $CoMo/Al_2O_3$.In the present study catalysts based on supported Ni and Mo are evaluated as catalysts for HDO reaction. The effect of metal phase and support properties, SBA-15 and γ-Al_2O_3 (reference support), as well as their interactions has been investigated. For the catalytic reactions, guaiacol was chosen as a model compound for lignin depolymerization fragments formed during bio-oil pyrolysis from lignocellulosic biomass.

2. **Methods**

The pure mesoporous silica SBA-15 was synthesized by hydrothermal method. The γ-Al_2O_3 was purchase from Degussa. The bimetallic nickel-molybdenum catalysts were prepared by wet impregnation. First, appropriate quantities of $(NH_4)_6Mo_7O_{24}.4H_2O$ and $Ni(NO_3)_2.6H_2O$ were dissolved in deionized water; the support was added and stirred for 2 h. Then, the suspension was freeze dried and the catalyst was calcined at 500 C for 2 h, under air flow. The metal contents were 15% of MoO_3 and 5% de NiO (weight percentage) with relation to support. The materials were characterized by nitrogen physisorption isotherms, transmission electron microscopy (TEM), X-ray powder diffraction (XRD), energy-dispersive spectroscopy (EDS) and with Raman and infrared spectroscopy. The total acidity of the catalysts was measured by Temperature programmed desorption of NH_3 (TPD-NH_3). The catalytic tests were carried out in a fixed bed flow reactor system under 15 kgf/cm^2 pressure. The catalysis, mass around 200 mg, was pre-treated in situ with 50 cm^3/min flow of pure H_2 at 500° C for 4 h and then cooled to the reaction temperature (200, 250 or 300° C). The feed mixture consisted of 2% w/w guaiacol (MERCK) in heptane P.A. (Aldrich) with 14.8

cm^3/h flow and 100 cm^3/min H_2 flow. The feed and the products were analyzed by gas chromatography Shimadzu CG17A with FID detector, capillary column CPsil 5CB and automatic injection.

3. Results and discussion

The fresh catalysts X-Ray diffractogram show that the Ni and Mo were successfully incorporated under the SBA-15 and Al_2O_3 supports as NiO and MoO_3 nanoparticles. After nickel and molybdenum incorporation the surface area of the SBA-15 decrease from 630 to 332 $m^2.g^{-1}$. The same behavior was observed for γ-Al_2O_3 support, the surface area chance from 279 to 233 $m^2.g^{-1}$. The temperature reduction of Ni and Mo oxide were determined by TPR-H_2, 500 °C under H_2 atmosphere for 4 hours. The catalysts total acidity for NiMo/SBA and NiMo/ALU were 470 and 548 µmol.g^{-1}, respectively. The results of catalytic activity for both materials are summarized in Table 1. For NiMo/SBA and NiMo/ALU the reaction temperature have a positive effect in guaiacol conversion, the higher results were observed at 300 °C 84.7% and 72.6% for NiMo/SBA and NiMo/ALU, respectively. However, the higher HDO values occurred for reaction temperature at 200°C. Thus, the selectivity for deoxygenated compounds were more pronounced in low temperatures for both catalysts. The principal products formed during the guaiacol HDO reaction were benzene, cyclohexane, cyclohexene, phenol and o-cresol. Compared to NiMo/ALU the NiMo/SBA showed the best performance in HDO, HDA and higher selectivity for benzene in all reaction temperature studied.

Table 1: The catalytic results of NiMo/SBA and NiMo/ALU in the hydrodeoxygenation of guaiacol.

	NiMo/SBA			NiMo/ALU		
Temperature (°C)	**200**	**250**	**300**	**200**	**250**	**300**
Conversion (%)	3.7	74.5	84.7	5.1	45.0	72.6
HDO (%)	92.1	74.3	66.5	37.9	31.5	30.8
HDA (%)	84.2	48.7	35.3	23.1	12.3	2.7
Selectivity (%)						
Benzene	16.7	8.6	1.3	3.3	1.1	0.3
Cyclohexane	29.1	13.9	11.0	6.8	3.3	0.4
Cyclohexene	38.5	26.2	19.5	12.9	7.9	1.9
Phenol	10.5	17.4	34.2	16.3	24.8	41.3
O-Cresol	0.0	18.4	8.0	7.7	8.5	11.3

With the catalysts thermogravimetric analysis were possible quantified the coke formation after HDO reactions. For NiMo/ALU and NiMo/SBA the coke formation were 1,65 and 1.14 $mg.g_{cat}^{-1}.h^{-1}$.

4. Conclusions

Compared with NiMo/ALU the NiMo/SBA show the best performance in guaicol conversion, selectivity for deoxigenated compound and in HDO percentage. The principal compounds formed during the HDO reaction with NiMo/SBA were cyclohexene, cyclohexane and bezene. For NiMo/ALU the principal compounds were phenol and o-cresol. Thus, the mesoporous silicon oxide (SBA-15) was the best support for nickel e molybdenum nanoparticles resulting in the catalyst for guaicol hydrodeoxygenation.

References

- Arun, R.V. Sharma, A.K. Dalai, Renew. Energy Rev., 48 (2015) 240-255.

Keywords

Hydrodeoxygenation, catalysis, nickel, molybdenum.

530385 Hydroprocessing of Biomass-Derived Oxygenates on Metal-Exchanged Zeolites Using Light Alkanes As the Source of Hydrogen

Abstract:

The broader use of natural gas and biomass for energy applications is limited because of the short hydrocarbon chains and/or high oxygen content of the molecules that comprise these resources. Current technologies for natural gas and biomass upgrading require high temperatures and suffer from poor selectivities. Furthermore, biomass deoxygenation requires external sources of H_2 molecules that are most commonly derived by sacrificing a portion of the feedstock in steam reforming reactions. In the absence of molecular hydrogen, a portion of the carbon feedstock is converted to arenes during hydrogen transfer reactions to satisfy the stoichiometry for oxygen removal as water. This work aims to improve selectivity and solve the hydrogen deficiency issue by developing new catalysts that can facilitate hydro-deoxygenation reactions using light alkanes as the hydrogen source. These hydrogen transfer steps have the potential to selectively terminate growing hydrocarbon chains via desorption as less reactive alkanes, thereby producing targeted classes of compounds, as opposed to a wide range of products over a broad carbon number distribution. Furthermore, this strategy has the potential to upgrade a portion of the light alkane feed because the hydrogen transfer steps generate alkenes that can participate in subsequent C-C bond formation reactions.

Reactions of isobutane and n-butanal (either as single reactants or co-feeds) on H-BEA zeolite and H-BEA samples partially exchanged with Zn cations were studied to probe the effects of alkanes as co-reactants during acid-catalyzed reaction of oxygenates. The Zn-exchanged BEA zeolite was synthesized via ion-exchange using aqueous solutions of zinc nitrate followed by treatment in flowing, dry air first at 373 K for 20 h then at 773 K for 4 h. This method has been previously shown to effectively exchange protons for Zn^{2+} ions within the cages of zeolites. n-Butanal reactions (2 kPa) on H-BEA predominantly form 2-ethyl-2-hexenal via aldol-condensation (60% selectivity to the aldol-condensation dimer) while also forming xylenes and cyclic ketones via cyclization/deoxygenation of 2-ethyl-2-hexenal (25% selectivity). n-Butene formation was also observed (selectivity of 15%) via hydro-deoxygenation of n-butanal. Addition of isobutane as a co-feed on H-BEA leads to minor changes in the product selectivity, indicating that the unsaturated species formed from n-butanal conversion are not strong enough hydride acceptors to activate isobutane via hydride transfer on Bronsted acid sites. Product selectivities for n-butanal conversion on Zn-H-BEA were similar to those on H-BEA, suggesting a minor effect of the Zn sites on n-butanal reaction paths. Co-reaction of isobutane (40 kPa) and n-butanal (2 kPa) on Zn-H-BEA exhibited a completely different product distribution. The predominant products were C_4 hydrocarbons (34% selectivity with 50% selectivity to n-butane within C_4) and C_8 alkanes (dimethyl-hexane and methyl-heptane isomers; 31% selectivity). n-Butane is the expected isomer formed from deoxygenation-hydrogenation of n-butanal, and the C_8 isomers that were produced were those that would be expected from deoxygenation of aldol-condensation (methyl-heptanes) and Prins-condensation (dimethyl-hexanes) products. Reaction of isobutane/n-butanal feeds on the Zn-H-BEA catalyst also exhibited lower selectivity to the aldol-condensation dimer (16% versus 70%) and to cyclic species (16% versus 25%) compared to isobutane/butanal reactions on H-BEA. Isopentane formation (5% selectivity) and C_7 alkanes (methyl-hexanes and n-heptanes; 9% selectivity) were also observed on Zn-H-BEA during conversion of isobutane/butanal feeds. These C_7 isomers are what would be expected from deoxygenation of C_7 ketones derived from Tishchenko esterification-ketonization routes. Reaction of isobutane (40 kPa) was also carried out on H-BEA and Zn-H-BEA to identify the extent of isobutane dehydro-oligomerization reaction paths. Isobutane conversion was only 0.1% on H-BEA and only 0.2% on Zn-H-BEA, indicating that dehydro-oligomerization of isobutane only occurs to a negligible extent in parallel with n-butanal conversion paths. This negligible conversion of isobutane is consistent with the absence of trimethyl-pentane isomers from the product distributions (indicating undetectable isobutane dehydro-oligomerization).

530346 Flowthrough Reductive Catalytic Fractionation of Biomass

Abstract:

As an abundant and renewable source of carbon, lignocellulosic biomass is an excellent feedstock for sustainable fuel and chemical production. Biomass primarily consists of carbohydrates (cellulose and hemicellulose) and lignin. Carbohydrates have been extensively studied and shown to be versatile feedstocks for the production of fuels and chemicals. Lignin, the largest source of natural aromatics, becomes highly recalcitrant during carbohydrate processing, hindering chemical production from lignin. Reductive catalytic fractionation (RCF) has emerged as a leading biomass fractionation and lignin valorization strategy that simultaneously extracts, depolymerizes and stabilizes lignin without disturbing the carbohydrate fractions. Here, flowthrough reactors were used to investigate RCF of poplar. Most RCF studies to date have been conducted in batch, but a flow-based process enables the acquisition of intrinsic kinetic and mechanistic data essential to accelerate the design, optimization, and scale-up of RCF processes. Time-resolved product distributions and yields obtained from experiments with different catalyst loadings were used to identify and deconvolute events during solvolysis and reductive stabilization. Multi-bed RCF experiments provided unique insights into catalyst deactivation, and initial rate studies enabled the determination of kinetic parameters intrinsic to the solvolytic step of RCF. Overall, this work demonstrates the concept of flowthrough RCF, which will be vital for realistic scale-up of this promising approach.

530547 Hydrodeoxygenation of Sorbitol to Monofunctional Fuel Precursors over Co/TiO_2

Abstract:

Over the next two decades, demands for heavier fuels such as jet and diesel are projected to increase while demands for lighter fuels such as gasoline are projected to decrease.[1] Production of these heavy C_8-C_{22} hydrocarbons remains a challenge when using biomass feedstocks in large part because the monomeric units which comprise cellulose and hemicellulose possess only five or six carbon atoms. One potential route for producing these fuels involves converting carbohydrates into C_{5+} monofunctional oxygenates (MFs) by aqueous-phase hydrodeoxygenation (APHDO) followed by performing C-C coupling reactions.[2] APHDO consists of feeding a highly-oxygenated species in water at moderate temperatures (~500 K) along with hydrogen at high pressures (~6 MPa) over a bifunctional catalyst usually utilizing a noble metal on an acidic support.[3] In this chemistry it is desired to control the cleavage of C-O bonds and minimize the cleavage of C-C bonds by retro-aldol condensation, decarbonylation and decarboxylation reactions.[4] However, over-dehydration to lower-demand alkanes is undesired. In addition, the noble metals commonly used are costly, but utilization of low-cost base metals is often limited in APHDO due to their propensity to leach into the aqueous phase. However, recent developments have shown that Co/TiO_2 catalysts can be stabilized for aqueous-phase hydrogenations utilizing high-temperature pretreatments which take advantage of the strong metal-support interaction (SMSI).[5] The purpose of this study is to investigate the potential for MF production in sorbitol APHDO using a SMSI-stabilized Co/TiO_2 catalyst.[6]

Sorbitol APHDO of Co/TiO_2 produced a 56% yield (on a carbon basis) to MFs at a moderate WHSV of 0.70 h^{-1}. This is the highest yield reported from sorbitol in the literature. We estimate the Co catalyst is 50 to 1,000 times cheaper than previously reported precious metal catalysts. Increasing space time leads to excessive deoxygenation and therefore selectivity toward alkanes and CO_x while decreasing space time leads to higher oxygenates, including diols, oxygenated heterocycles, unidentified O_{3+} species. The MFs produced are mostly primary alcohols (37% at 0.70 h^{-1}), secondary alcohols (32%), and heterocycles (23%). Ketones (7%) and aldehydes (<1%) are produced in much lower yields. 67% of the MFs are C_{5+} and would therefore be suitable for coupling. FTICR-MS shows that sorbitol/anhydrosorbitol species can oligomerize at lower contact times to form C_9-C_{18} species which appear to break down to some degree at longer contact times to eventually form MFs. Overall, the product distribution vs. WHSV suggests that about one third of the fed carbon reacts through the retro-aldol route and overall 20% of the carbon is lost to C_1 gases such as CO_2 and methane. The need to suppress retro-aldol condensation and decarbonylation in sorbitol APHDO is a clear limitation for the production of C_{5+} MFs. In the present study, Co/TiO_2 catalyst shows irreversible deactivation with TOS due to sintering and leaching of the Co. Overall, this work shows that Co/TiO_2 can be used to produce monofunctional fuel precursors from sorbitol at high yields, but suffers from irreversible leaching and sintering due to chelation from highly-oxygenated species.

References

1. ExxonMobil, 2017 Outlook for energy: A view to 2040. ExxonMobil: Irving, 2017.
2. Kunkes, E. L.; Simonetti, D. A.; West, R. M.; Serrano-Ruiz, J. C.; Gärtner, C. A.; Dumesic, J. A., Catalytic conversion of biomass to monofunctional hydrocarbons and targeted liquid-fuel classes. *Science* **2008,** *322* (5900), 417-421.
3. Vilcocq, L.; Cabiac, A.; Especel, C.; Guillon, E.; Duprez, D., Transformation of sorbitol to biofuels by heterogeneous catalysis: Chemical and industrial considerations. *Oil & Gas Science and Technology– Revue d'IFP Energies nouvelles* **2013,** *68* (5), 841-860.
4. Li, N.; Huber, G. W., Aqueous-phase hydrodeoxygenation of sorbitol with Pt/SiO_2–Al_2O_3: Identification of reaction intermediates. *Journal of Catalysis* **2010,** *270* (1), 48-59.
5. Lee, J.; Burt, S. P.; Carrero, C. A.; Alba-Rubio, A. C.; Ro, I.; O'Neill, B. J.; Kim, H. J.; Jackson, D. H.; Kuech, T. F.; Hermans, I.; Dumesic, J. A.; Huber, G. W., Stabilizing cobalt catalysts for aqueous-phase reactions by strong metal-support interaction. *Journal of Catalysis* **2015,** *330*, 19-27.
6. Eagan, N. M.; Chada, J. P.; Wittrig, A. M.; Buchanan, J. S.; Dumesic, J. A.; Huber, G. W., Hydrodeoxygenation of Sorbitol to Monofunctional Fuel Precursors over Co/TiO2. *Joule* **2017,** *1* (1), 178-199.

Numerical Simulation of Natural Fractures on EOR by Cyclic Gas Injection in Unconventional Reservoirs Using DFN Approach

Olatoyosi Obilade, Chongwei Xiao[1]
Texas A&M University - Kingsville

Abstract

Major published works on EOR by cyclic gas injection in shale reservoirs have focused on the use of Dual porosity/Dual permeability approach for fracture model development. To adequately account for fluid transport, storage mechanisms, and complex reservoir heterogeneities, natural fractures are modeled using the Discrete Fracture Network (DFN) modeling approach in this study. This allows for discrete modeling of the natural fractures as against an equidistant approach and improves the accuracy of 3D numerical model.

A 3D reservoir simulation model was developed for the enhanced oil recovery by cyclic gas injection in a shale reservoir using a commercial simulator. The field 3-D simulation of EOR by cyclic gas injection was carried out using a field in Zavalla County of the Eagle ford shale play. Fracture effectiveness based on fracture contribution to flow was considered in fracture model generation. Using the model the impact of natural fractures is evaluated. The effectiveness of DFN in fractured shale simulation was simulated with cyclic gas injection. The effects of fracture density and orientation of natural fractures, soaking period on ultimate EOR by cyclic gas injection were analyzed. The main driving mechanisms are the oil viscosity reduction and volume swelling. Compared to the conventional Dual porosity/Dual permeability model, the DFN modeling provides a realistic method of evaluating the potential recovery from EOR in shale reservoirs and moves us a step closer to successful field implementation of cyclic gas EORs in shale plays.

Key words: natural fracture, unconventional reservoir, cyclic gas injection, shale gas

Introduction

Many approaches have been used for improving/enhancing oil recovery in conventional reservoirs. These include water flooding (secondary recovery), gas miscible recovery, chemical flooding and thermal recovery. The study of the feasibility/viability of each of these methods as regards unconventional reservoirs will be critical in choosing an appropriate stimulation method for shale reservoirs. Water flooding and chemical flooding in shale reservoirs do not appear to be feasible improved recovery techniques because of the tight matrix causing low injectivity issues. Furthermore, many shale formations are made up of clay minerals which can cause severe permeability alterations during water flooding(Lan, Dehghanpour et al. 2014). Chemical flooding can also result in the alteration of the properties of the shale and faces the challenge of chemical adsorption for surfactant and retention for polymer. Thermal methods are usually applied to heavy oil. Since oil viscosities in shale reservoirs are usually low, the benefit from this method cannot be significant(Sheng 2015). Gas injection solves many of the challenges faced by other EOR techniques in unconventional reservoir. It is a fast and clean method. In CO_2 gas injection, the CO_2 is injected in dense supercritical condition, the oil vaporizes and flows out as gas. Gas injection can either be in gas flooding mode or cyclic gas injection mode. EOR in naturally fractured reservoir by huff-n-puff gas injection has been studied through experimental work (Gamadi, et al 2013;Gamadi, et al. 2014), and simulation work (Chen et al. 2013; Wan et al. 2013; Wan et al. 2014; Wei et al.2014). The results of the aforementioned research work show that huff-n-puff gas injection can effectively improve

[1]: The corresponding author, Chongwei.Xiao@tamuk.edu, +1 3615932746

recovery in naturally fractured shale play. Huff-n-Puff gas injection three phases: Phase one is gas injecting where CO_2 is injected into a reservoir using the same producing well as against using a separate well for injection; In phase 2, the well is shut in at the bottom or at the well head to push the injected fluid further into the reservoir formation; In Phase 3 the well is put back into production. There are a number of reasons why the huff-n-puff gas injection is seen as effective choice compared to conventional gas flooding. It does not require the drilling of a secondary well for injection, which makes it a lower risk process and avoids well to well connectivity issues that might be faced in conventional gas flooding and breakthrough of injected fluids in producing wells(Kong, Wang et al. 2016). The complex system of hydraulic and preexisting fractures provides a pathway for produced oil and injected gas and also sufficient contact area for the injected gas for penetration and diffusion into the low-permeability matrix. The aim of all EOR methods is to increase the volumetric (macroscopic) sweep efficiency and to improve the displacement (microscopic) efficiency(Mathiassen 2003). The volumetric sweep can be increased by reducing the mobility ratio between the displacing and displaced fluids. The microscopic efficiency is improved through reduction of the interfacial tension between injected solvent and reservoir fluids which reduces the amount of oil trapped due to capillary effects. Although conventional CO_2 flooding has an excellent microscopic displacement efficiency when multiple contact miscibility is achieved, poor sweep (macroscopic) efficiency may be obtained because of the channeling of CO_2 through high-permeability fracture connectivity pathways between the injection and production wells (Sanchez-Rivera, Mohanty et al. 2015). Therefore, the CO_2 Huff-and-Puff process is a preferred approach for these reservoirs which have both complex high permeability fracture networks and ultra-low permeability matrix.

Natural Fractures in Oil Reservoirs

Natural fracture presence in shale reservoirs have been researched extensively to identify their properties and prove their existence. Gale (2014) carried out extensive research on natural fractures in shale reservoirs by studying rock outcrops. The intensity of bedding parallel fractures in various shale plays within the United States was shown in Figure 1. The distinction between natural and induced fractures being that the former is produced by drilling, hydraulic fracturing, bore-hole investigations or other human related activities (Kulander, Dean et al. 1990). Natural fractures can also be open, partially closed or closed (Zangeneh, Eberhardt et al. 2014). However closed natural fractures can be reactivated or stimulated by hydraulic fracturing in horizontal wells (Fisher, Heinze et al. 2004).

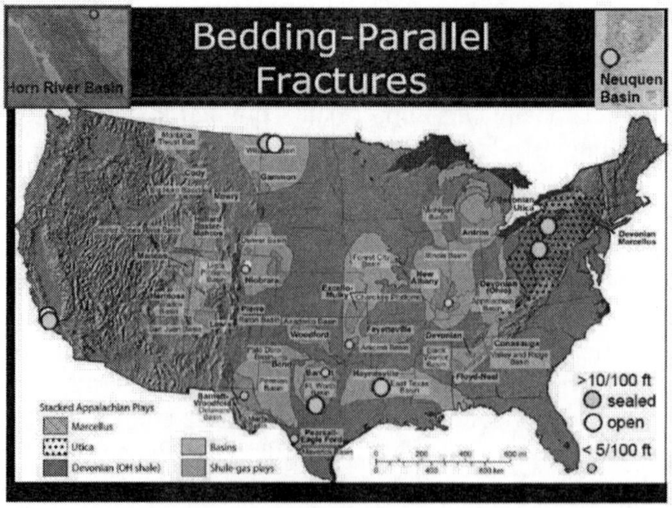

Figure 1. Intensity of natural fractures in some US shale plays (Gale 2014)

The presence of natural fractures in a shale reservoir significantly affects fluid flow in a reservoir (Gilman 2003). The role of natural fractures and effects of natural fracture parameters on production in unconventional reservoirs have become increasingly investigated by oil and gas researchers. A presumption on oil production in shale formation is that "because the formations are so tight, commercial production only happens when extensive networks of natural fractures exist" (Carlson and Mercer 1991). Natural fracture presence is one of the most important considerations in defining the economic prospect of a shale formation. Theloy, Sonnenberg et al. (2011), while studying the effects of natural fractures on production in the Sanish field of North Dakota, concluded that the natural fracture presence is a reasonable explanation for an increased production figures obtained in the eastern part of the field. Also, field production variability (FVP) plots which relate the fractional cumulative produced hydrocarbon and the fractional cumulative number of wells have been used to show that fracture density in wells is directly responsible for the variation in production amongst wells drilled in similar formations(Gonzalez and Aguilera 2011, Lopez, Gonzalez et al. 2012). Natural fractures in the low permeability formations make significant contribution to production, and therefore it is pertinent to investigate their characteristics (Bahrami, Rezaee et al. 2012). This is because the natural fractures create a layer drainage volume than would be obtainable from just hydraulic fracturing. Therefore, identifying natural fractures characteristics in the low permeability formations becomes important for well productivity evaluations. Furthermore, the connectivity of natural fractures in a reservoir have a very significant effect on the response to fluid injection including hydraulic fracture propagation (Riahi and Damjanac 2013, Zangeneh, Eberhardt et al. 2015).

An understanding of the storage and transport mechanisms in shale reservoir is required for optimizing development of shale reserves. A prevailing approach is to modeling fractured reservoirs by the dual porosity/dual permeability method. This approach is designed to simplify reservoir simulation for shale reservoirs using the dual-continuum concept in which fracture and matrix are modeled as two separate kinds of continua occupying the same volume in space (Warren and Root 1963). The fractured reservoir is represented by two combined systems, the matrix with low permeability and the fractures with high permeability. Fluid flow is assumed to take place separately in the fracture and matrix with some fluid interchange at the interface with the properties of the matrix and fracture being averaged properties of the medium. Therefore, the averaging volume should be greater than the REV (Representative Elementary Volume) of the porous systems. Finding an appropriate REV however is a challenging process as shown by Leung (2009). The double porosity/dual permeability model is simple and needs few parameters to characterize the fractured systems. Whereas, this approach does not explicitly model the geometry and the discrete features of natural fractures. It gives a decent estimation of the flow in shale reservoirs but it's generally too simple to capture the influence of complex fracture networks (Anupam 2010). This inaccurate model of fracture connectivity results in inaccurate flow predictions. This leads to the application of discrete fracture network (DFN) models as an alternative approach for fractured reservoir modeling. DFN incorporates the geometry and individual characteristics of discrete features as a primary factor affecting fluid flow and fluid transport. DFN models can be deterministic or stochastic. Stochastic DFN implementations incorporate statistical distributions from analysis of fracture parameters (density, length, spacing, orientation, and aperture size) (Jin and Pashin 2008). This approach results in a more accurate representation of the reservoir and therefore produces reservoir models that give more realistic predictions about flow in the reservoir. In the DFN fracture representation, fractures open and contributing to flow are explicitly presented as 3-D polygons. Matrix-fracture coupling coefficients and other properties are then applied to each polygon. The advanced model, Discrete Fracture Networks (DFN), models the fractures discretely/explicitly and therefore improves the accuracy of modeling the complex system of natural fractures. The main disadvantage is in increased computational resource requirement(Wu, Di et al. 2011), especially with increasing domain size. To authors' best knowledge, no work has been published to model improved recovery through huff-n-puff in shale reservoirs using DFN modeling for explicit fracture representation.

This study developed a simulation model representing an unconventional reservoir with natural fractures using the DFN approach. Different natural fracture systems were defined by discrete fracture network model (DFN) with different fracture orientation, intensity, and length. The mechanisms of oil recovery by huff-n-puff gas injection under different operation conditions and soaking periods by compositional simulation were investigated.

Eagle Ford Shale

The Eagle Ford shale (EFS) play stretches from the Maverick basin in Southern Texas to the Eastern Texas basin (http://info.drillinginfo.com/eagle-ford-geology-assessment-and-activity-update/). The Eagle Ford shale lies above the Buda limestone and is overlain by the Austin Chalk. The EFS has long been regarded as a shale source rock but recently it is being exploited as a producing formation. Reactivated faults in the Eagle Ford shale provided pathways and barriers to up-dip migration. In some down dip areas, they acted as barriers giving rise to huge hydrocarbon accumulations, but in other areas, acted as pathways to allow the oil to migrate up-dip (Condon and Dyman 2003).

The thickness of the Eagle Ford ranges from 120 to 500 ft and depth ranges between 2,500 to 14,000 ft. Initial reservoir pressure also differs in parts of the shale play with over-pressuring seen in areas mostly down-dip of the Edwards reef trend (Mullen 2010). Developing the Eagle Ford shale faces numerous challenges including: a variation in the shale properties across the Eagle Ford; a variation in fluids (dry gas, wet gas, oil) produced from different areas; some regions have natural fractures while natural fractures are absent in other regions, which affects productivity, reserves estimation and production forecasting in the Eagle Ford; the extremely low permeability in the Eagle Ford which yields long transient-flow periods (Gong 2013), etc. Furthermore, the Eagle Ford shale has high carbonate content and a low clay content, which makes it more brittle and easier to produce using hydraulically fractured wells (Pope, Palisch et al. 2012).

Hydraulic fracturing and horizontal wells are credited as the most effective technologies for improving the productivity of unconventional reservoirs. Hydraulic fractures reduce formation damage and increase the conductivity of path of fluid flow to the wellbore. Hydraulic fracturing has been shown to be appropriate for well stimulation in various kinds of unconventional reservoirs.

A major driving force for CO_2 enhanced oil recovery (Alexander, Baihly et al.) in shale reservoirs is pressure maintenance (Hawthorne, Gorecki et al. 2013, Sheng and Chen 2014). Also, the oil mixes with the gas which results in lower viscosity and higher mobility (Shelton and Morris 1973). The dissolution and diffusion of CO_2 in crude oil also causes oil swelling. The altering of wettability and relatively permeability hysteresis is also another important mechanism (Denoyelle and Lemonnier 1987, Haines and Monger 1990). During the gas injection phase, with the gas phase being the continuous phase, mobility is instant. During the production phase however, the gas does not begin to flow until critical gas saturation is reached. Summarily, at the same gas saturation relative permeability of gas is higher during the injection phase than during the production phase. The reduction in interfacial tension (Kong, Wang et al. 2016) and improved sweep efficiency (Sheng 2015) are also mechanisms contributing to improved oil recovery in unconventional reservoirs. The huff-n-puff gas injection can also lead to re-fracturing in shale reservoirs that will improve the recovery efficiency (Vincent 2011).

Huff-n-puff in Unconventional Reservoirs

Because EOR in shale reservoir is a relatively new field of research, there is limited work to present. Wang , Sheng and Meng Wan (2013), Wan, Sheng et al. (2013), Wan, Sheng et al. (2013), (Wan, Meng et al. 2014) pioneered simulation of huff and puff gas injection in unconventional fractured reservoirs. The study concluded that cycle time affects the oil recovery. The study also examined the effect of fracture spacing vis-a-vis fracture conductivity on huff and puff performance. The simulation results postulated that the fracture spacing is a more significant factor than fracture conductivity. Also the effects of planar oriented hydraulic fractures on enhanced oil recovery from shale reservoir were investigated. Findings showed that total oil recovery from a fractured oil reservoir increased from 6.5% for primary depletion to up to 40% with full miscibility and higher cycle frequency. However the study did not investigate the effects of soaking period on final oil recoveries. Furthermore, compositional model was used to examine the effect of the composition of injected gas on cyclic gas injection. As expected, recovery increases with the richness of the gas. Gamadi, Elldakli et al. (2014) also carried out a compositional simulation of huff-n-puff natural gas injection as against huff-n-puff CO_2 and investigated how soaking period, cycle periods and injection pressures affect ultimate recovery. Chen, Balhoff et al. (2014) compared the recovery by CO_2 huff-n-puff in a homogeneous reservoir to recovery from a heterogeneous reservoir and indicated that there was a linear correlation between reservoir heterogeneity and recovery rate.

Model Development and Description

To study the performance of huff-n-puff CO_2 injection, a 3-D reservoir model was developed with a well located in the Zavala County which is mainly in the oil window of the Eagle Ford shale play. Reservoir properties were obtained from published data based on the Eagle Ford shale, well logs, drilling and completion reports (Gamadi et al. 2014; Railroad Commission 2017; Wan et al. 2013b). The permeability and porosity of the Eagle Ford reservoir is chosen to be 0.0003md and 8% respectively which are comparable to values from earlier research work.

Due to the limited fluid sampling activity in the Eagle Ford shale play, fluid PVT information is relatively difficult to obtain (Orangi et al. 2011). A typical Eagle Ford fluid PVT data obtained from Gamadi et al. (2014) was used in this study. To confirm the accuracy of the information, the composition of the fluid was compared with PVT fluid data from the Eagle Ford shale in Phi and Schechter (2017). Two sets of relative permeability curves are used for natural fractures and the matrix, respectively. Reservoir rock properties are assumed to be homogeneous and isotropic.

A 40 acre drainage density is considerable reasonable for an oil well in this formation (Chandler 2010). The shale oil reservoir model dimension is 6000 ft. in length x 1000 ft. width x200 ft. thickness. Height of the pay zone is 200 ft. Initial reservoir model is 20*20*10 (15ft*50ft*20ft) grid blocks prior to local grid refinement. However, as discussed earlier, to limit the computational complexity of the model, the SRV of a single hydraulic fracture is simulated. This gives rise to a simulation model of 20*20*10 (15ft*50ft*20ft) grid blocks prior to the local grid refinement. The initial reservoir pressure is 4800 psi with a minimum bottom hole pressure (BHP) constraint of 2500 psi. The bubble point pressure of the reservoir fluid is 2430 psi. Implicit assumption is that the reservoir is always above bubble point pressure to avoid the complex calculation associated with multiphase reservoirs. The production is therefore BHP controlled keeping the reservoir pressure above bubble point pressure to prevent gas liberation at reservoir conditions.

The logarithmically spaced, locally refined and dual permeability (LS-LR-DK) method was applied to simulate fluid flow in the hydraulic fractured oil shale reservoirs. The dual-permeability method allows us to model fluid communication between the fracture and matrix as well as between the matrices. This however creates increased computational difficulty as evidenced by increased simulation time (Uleberg and Kleppe 1996). This difficulty can be ascribed to the huge differences in the permeability of hydraulic fractures and matrices and further compounded by the need of this research work to explicitly implement natural fractures in this reservoir.

The field is developed with a horizontal well of lateral length 5400 ft. Total well length is 12287 (MD). Mendoza et al. (2011) created a detailed petrophysical analysis in a formation with similar rock and field properties and concluded that transverse hydraulic fractures are anticipated based on the direction of the well bore. Furthermore, the study included the analysis of the optimum spacing of the natural fracture on the expected net present value (NPV) for a 3800 ft. lateral horizontal well length was 200 ft. apart. As the optimum spacing was dependent on the lateral well length, we have chosen a 300 ft. spacing for our horizontal well resulting in 18 transverse hydraulic fractures because well completion information regarding hydraulic fracture spacing is not available. The fracture half-length is 250 ft which is shorter than real propped fracture half-length (Cipolla et al. 2009) with an average fracture width is 0.25 inch (0.006m) (Gokaraju 2014; Perkins and Kern 1961). Fracture is assumed to be confined to the pay zone and a height of 200ft stretches from top to bottom covering two-thirds of the pay zone with length skewed towards the upper portion of the fracture perforation. With a fracture half-length of 250ft, 18 stages, simple (bi-wing) fracture with 200ft length, we have a 1,800,000 ft^2 contact area which is approximately equal to the 40-acre density for oil well in the Eagle Ford region (Chandler 2010). LGR is used in the area surrounding the fractures in order to adequately capture flow representation and pressure drop around the fractures.

Sensitivity Studies based on SRV

The factors that affect recovery from primary production and secondary production by gas injection can be broadly classified into two groups:

- The intrinsic reservoir properties which are usually fixed and cannot be easily altered but which are however important in investment and operational decision making process in selecting potential reservoirs for primary and secondary recoveries.
- The operational factors which include the well scheduling and other production factors (such as bottom hole pressure (bhp), injection pressures, huff-n-puff well shut-in period, huff-n-puff injection period, huff-n-puff production period) that can be varied and affected by the operator.

In this work, the effects of the reservoir properties, specifically the effects of the natural fracture parameters on the productivity of shale reservoirs during the primary production phase and during improved recovery by the huff-n-puff CO_2 injection were investigated regarding following natural fracture parameters: stochastic seed constant in natural fracture generation, fracture orientation; fracture Intensity, and Fracture length.

1. Effect of Seed Constant

The seed number is used for initializing the random number generation process for the natural fracture system. In order to investigate the effect of the seed constant, the initial seed constant is varied from the base value of 32351 to a value four times that (129404) while keeping all the other fractal parameters the same. Figure 2 shows a small area (near-well) comparison of having a fracture network distribution based on seed constants of 32351 and 129404. For the 32351 seed constant, fractures are closer to the wellbore.

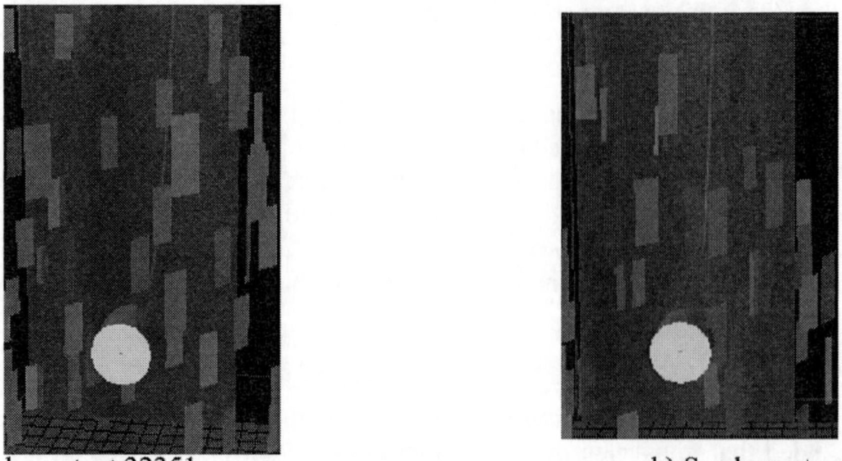

a) Seed constant 32351 b) Seed constant 129404

Figure 2. Fracture distribution near well bore

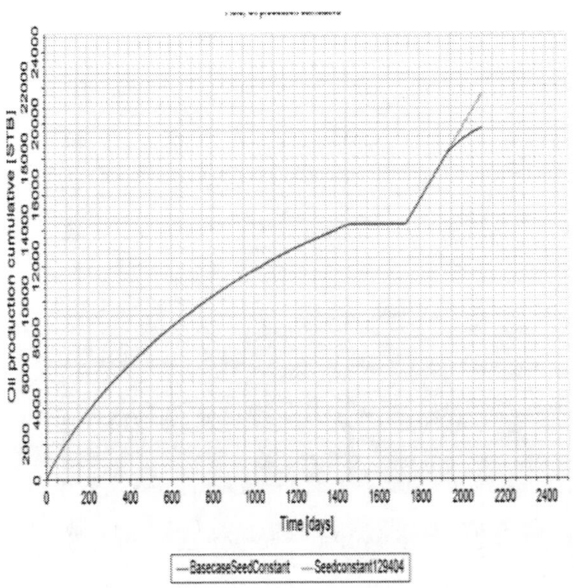

a) Cumulative oil production

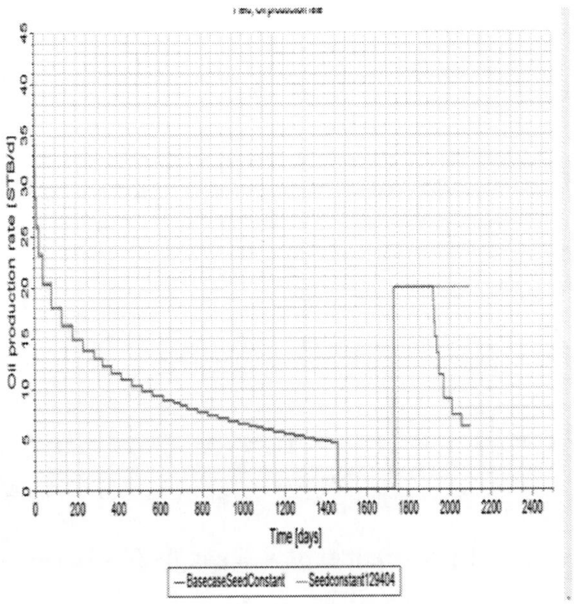

b) Oil flow rate

Figure 3. Effect of varying seed constant

In the early time, the response of cumulative production and oil production rates are similar. The reason is production at the early stage is mainly from the hydraulic fractures. However, as time progresses there is a divergence in oil production rate and cumulative production with the base case experiencing a higher production rate and cumulative production as shown in Figure 3. It is clear that while early time production is governed by the hydraulic fractures, production in later time is more controlled by the natural fractures.

2. Effect of Fracture Orientation

Fracture orientation has a huge impact on the productivity of shale reservoirs. Fracture orientation is made up of dip and azimuth. We examine the impact of various dip angles and fracture azimuth on productivity in both the primary and improved recovery production scenarios. In the 90 o dip angle, the natural fractures are vertically oriented in the same hydraulic fractures while in the 0 o they are oriented in the normal direction to the hydraulic fractures and the well bore. For the 45 o dip angle, there is an increased intersection between the natural and hydraulic fractures. Figures 4 shows the steorenet plots and close wellbore diagrams showing the intersections.

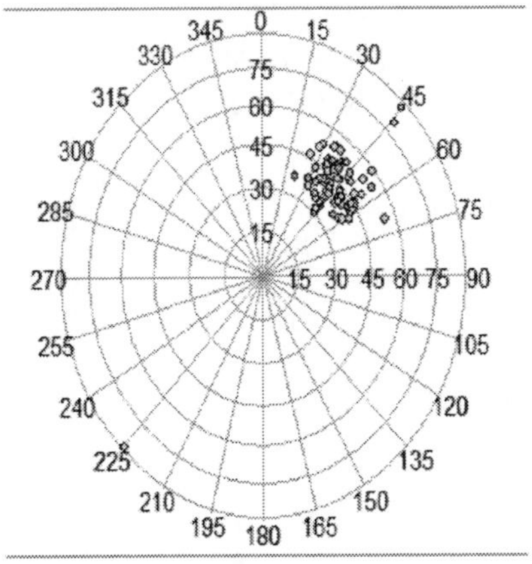

a1) Stereonet plot dip angle 45

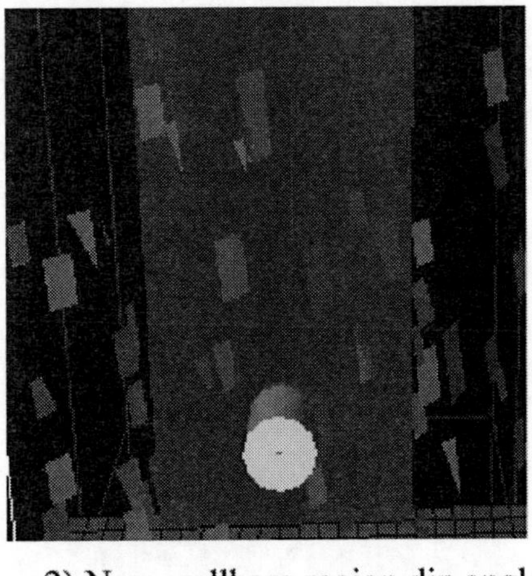

a2) Near wellbore region dip angle 45

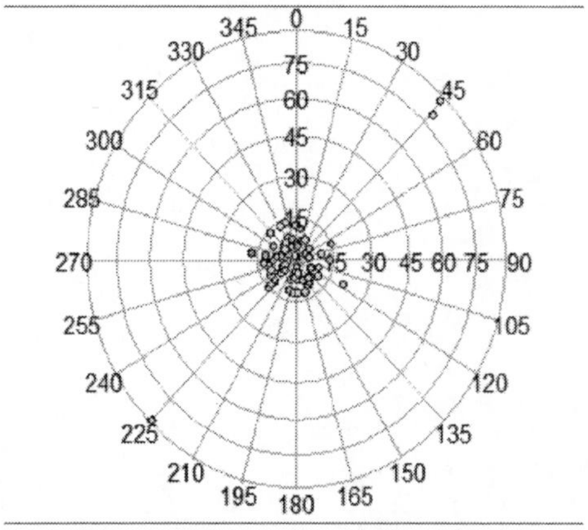

b1) Stereonet plot dip angle 0

b2) Near wellbore region dip angle 0

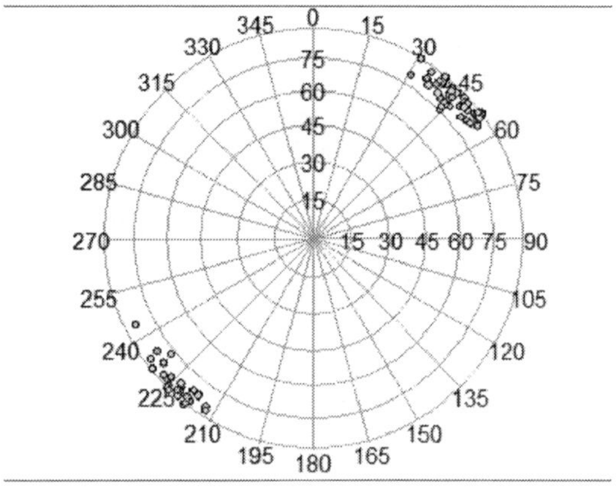

c1) Steorenet plot dip angle 90

c2) Near wellbore dip angle 90

Figure 4. Fracture orientation distribution

Sensitivity analysis was carried out for dip angles of 0°, 45°, and 90° (base case). For the azimuth, sensitivity analysis was carried out for azimuth 0°, 45° (base case) and 90°. In all the simulation runs, the orientation follows a Fisher model pdf. Results are shown in Figure 5.

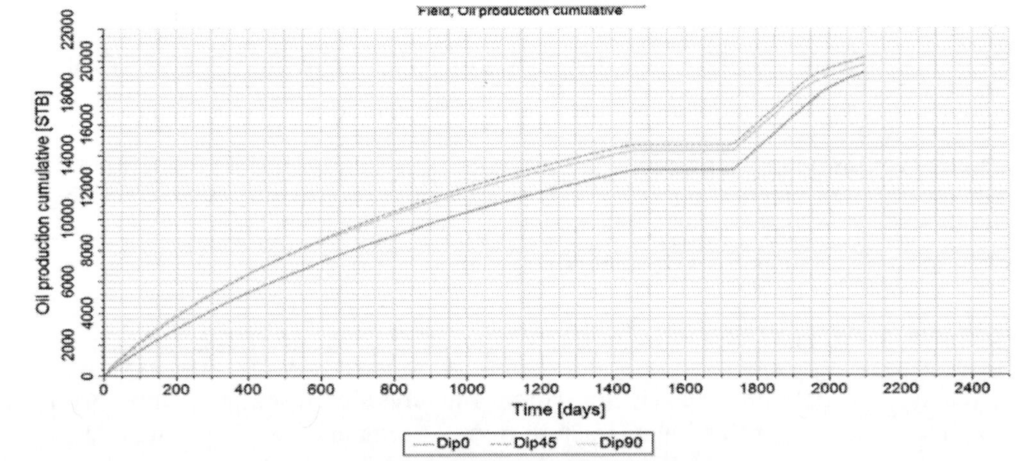

a) Cumulative oil production

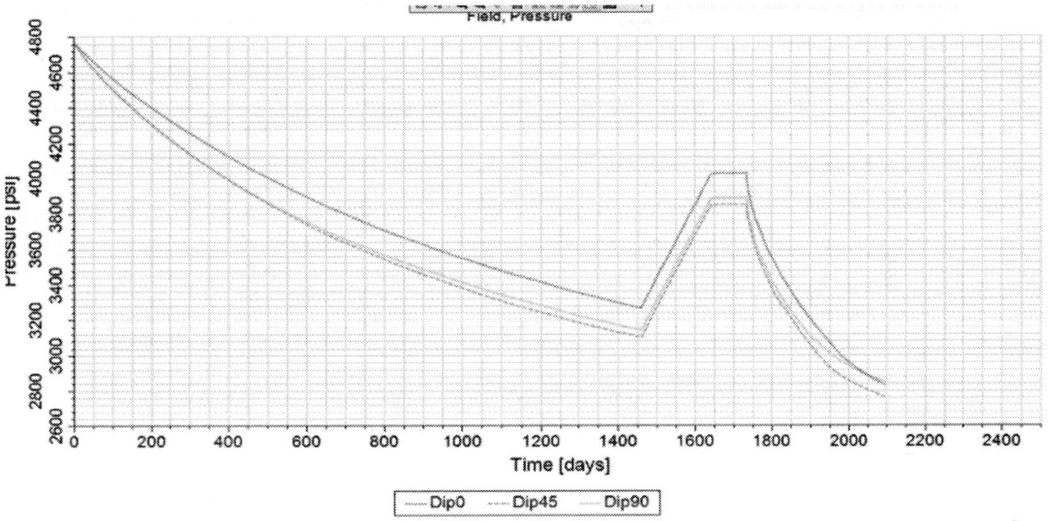

b) Oil production rate

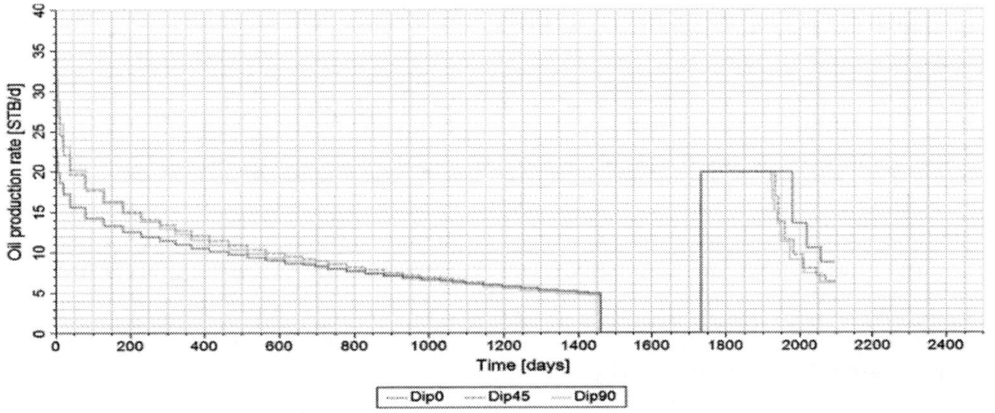

c) Average reservoir pressure

Figure 5. Effects of natural fracture dip angle

The dip angle of 0° shows a minimum oil cumulative production. The dip angle of 45° shows a slightly higher cumulative production figure than the 90° dip angle. There is a higher deviation between the cumulative production figures for 0° to 45° compared to the productions from 45° to 90°. As the dip angle nears 90°, the effect of dip angle variation is less appreciable in the primary production phase. However during the huff-n-puff CO_2 injection phase there is a larger deviation between the 45° and 90° dip angle even though they have a similar response during primary production. This same trend is observed in the well production rate and the field pressure decline.

3. Effect of Natural Fracture Length

Fracture length and shape determine the size of natural fractures. In all our simulation, we have kept our fracture shape the same. Therefore the fracture length remains the key factor in fracture size determination. The fracture size which is directly related to the fracture length affects the fracture connectivity and therefore the overall permeability of the system. Longer fracture lengths result in larger fracture sizes which result in higher fracture connectivity and real permeability and therefore higher productivity. Natural fracture lengths are varied between 25ft, 50ft (base case) and 100ft. The results are presented in Figure 6.

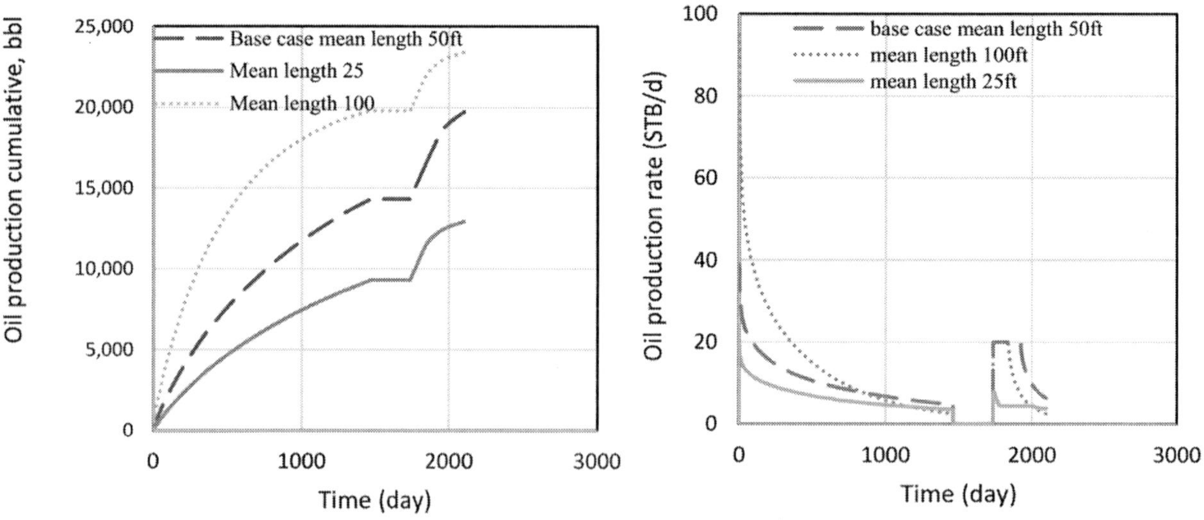

a) Cumulative oil production

b) Oil production rate

Figure 6. Effect of natural fracture length

The oil cumulative production and production rates are the highest for the 100ft natural fractures and the lowest for the 25ft natural fractures. In the huff-n-puff improved recovery, this linear proportion between the natural fracture length and the productivity is retained. Another key aspect of the natural fracture length investigated is the impact of the standard deviation from the mean length on flow from fractured reservoirs. The standard deviation is a measure of dispersion or variation from a central value. Low standard deviation implies the data points are close to our mean and vice versa. From our literature review, we have assumed a log-normal distribution for natural fracture length in the stochastic generation of fractures. This means that the logarithm of the fracture length is a normal distribution with a standard deviation of σ.

Conclusion and future work

A combination of log-generated and published field and fluid information were used in the model generation. The application of DFN model for natural fractures and dual permeability dual porosity model can well represent reservoir performance and increase the accuracy of reservoir simulation for the development of unconventional reservoirs. The seed constant/number affects the natural fracture distribution but does not affect reservoir performance in the primary recovery phase. The seed constant has more effect on fluid flow during the secondary production from fractured reservoirs with increased huff-n-puff cycles. The orientation (dip angle and azimuth angle) of natural fractures affects production in the primary and secondary recovery phases. A dip and azimuth angle of 45o for natural fractures shows the highest productivity. Minimal deviation in primary production figures between the 45o and 90o as compared to huff-n-puff phase. Variation in the azimuth angle has a more profound effect on production in both the primary and secondary recovery phases than varying the dip angles. There is a linear correlation between the length of natural fractures and productivity. A standard deviation shift of 5% - 20% from the mean length does not result in significant change in reservoir performance in the primary or secondary recovery phases.

It is important to note that the properties of natural fractures exist in any given reservoir. Recognizing their effects on primary and secondary recovery will allow the operator to make better decisions regarding investment, well planning, well location, well spacing and selection of appropriate reservoir for secondary huff-n-puff stimulation. Simulation of more huff-n-puff cycles can be carried out to examine the effect of production factors like soaking period, injection period, and production period on ultimate recovery from fractured reservoirs.

Acknowledgement

We greatly appreciate the support from Schlumberger for their generous donation of the software.

References

Alexander, T., J. Baihly, C. Boyer, B. Clark, G. Waters, J. Le, C. Houston, R. Lewis, C. M. K. Dallas, J. Thaeler, B. E. Toelle, D. Bentley, P. Warsaw, D. Friend, J. Hresko, R. Mitchell, B. Sylvester, S. Thomson, L. Hogarth and Houston, 2011. Shale gas revolution. https://www.slb.com/~/media/Files/resources/oilfield_review/ors11/aut11/04_shale_gas_revolution.pdf.

Anupam, A., 2010. Hierarchical Modeling of Fractures for Naturally Fractured Reservoirs Masters.

Bahrami, H., R. Rezaee and M. Hossain, 2012. Characterizing natural fractures productivity in tight gas reservoirs. Journal of Petroleum Exploration and Production Technology 2(2): 107-115.

Carlson, E. S. and J. C. Mercer, 1991. Devonian Shale Gas Production: Mechanisms and Simple Models.

Chen, C., M. T. Balhoff and K. K. Mohanty, 2014. Effect of Reservoir Heterogeneity on Primary Recovery and CO_2 Huff 'n' Puff Recovery in Shale-Oil Reservoirs, SPE-164553-PA.

Condon, S. M. and T. S. Dyman, 2003. DDS-69-H NOGA, petroleum systems and geologic assessment Undiscovered Oil&Gas, W. Gulf, TX.

Denoyelle, L. C. and P. Lemonnier, 1987. Simulation Of CO_2 Huff 'n' Puff Using Relative Permeability Hysteresis.

Fisher, M. K., J. R. Heinze, C. D. Harris, B. M. Davidson, C. A. Wright and K. P. Dunn, 2004. Optimizing Horizontal Completion Techniques in the Barnett Shale Using Microseismic Fracture Mapping.

Gale, J. F. W., 2014. Natural Fractures in Shale Hydrocarbon Reservoirs. Search and Discovery(#41487).

Gamadi, T. D., F. Elldakli and J. J. Sheng, 2014. Compositional Simulation Evaluation of EOR Potential in Shale Oil Reservoirs by Cyclic Natural Gas Injection, SPE-2014-1922690-MS.

Gilman, J. R., 2003. Practical Aspects of Simulation of Fractured Reservoirs. International forum on reservoir simulation, Baden-Baden, Germany

Gong, X., 2013. Assessment of Eagle Ford Shale Oil and Gas Resources. Phd.

Gonzalez, L. and R. Aguilera, 2011. Effect of Natural Fracture Density on Production Variability of Individual Wells in the Tight Gas Nikanassin Formation
Haines, H. K. and T. G. Monger, 1990. A Laboratory Study Of Natural Gas Huff `N' Puff
Hawthorne, S. B., C. D. Gorecki, J. A. Sorensen, E. N. Steadman, J. A. Harju and S. Melzer, 2013. Hydrocarbon Mobilization Mechanisms from Upper, Middle, and Lower Bakken Reservoir Rocks Exposed to CO
Jin, G. and J. C. Pashin, 2008. Discrete Fracture Network Models Of The Secarb Carbon Sequestration Test Site, Deerlick Creek Field, Black Warrior Basin, Alabama.
Kong, B., S. Wang and S. Chen, 2016. Simulation and Optimization of CO_2 Huff-and-Puff Processes in Tight Oil Reservoirs
Kulander, B. R., S. L. Dean and B. J. Ward (1990). Fractured Core Analysis.
Lan, Q., H. Dehghanpour, J. Wood and H. Sanei, 2014. Wettability of the Montney Tight Gas Formation
Leung, J., 2009. Reservoir Modeling Accounting for Scale-up of Heterogeneity and Transport Processes. PHD.
Lopez, B. A., L. Gonzalez, R. Aguilera and F. Garcia-Hernandez, 2012. Effect of Natural Fracture Properties on Production Variability of Individual Wells in Multiphase Oil Reservoirs
Mathiassen, O. M., 2003. CO_2 as Injection Gas for Enhanced Oil Recovery and Estimation of the Potential on the Norwegian Continental Shelf, NTNU – Norwegian University of Science and Technology
Mullen, J., 2010. Petrophysical Characterization of the Eagle Ford Shale in South Texas
Pope, C. D., T. Palisch and P. Saldungaray, 2012. Improving Completion And Stimulation Effectiveness In Unconventional Reservoirs- Field Results In The Eagle Ford Shale Of North America
Riahi, A. and B. Damjanac, 2013. Numerical Study of Interaction Between Hydraulic Fracture and Discrete Fracture Network, ISRM-ICHF-2013-037.
Sanchez-Rivera, D., K. Mohanty and M. Balhoff, 2015. Reservoir simulation and optimization of Huff-and-Puff operations in the Bakken Shale. Fuel 147: 82-94.
Shelton, J. L. and E. E. Morris, 1973. Cyclic injection of Rich Gas into producing wells to increase rates from viscous-oil reservoirs. Journal of Petroleum Technology 25(08): 890-896.
Sheng, J. J., 2015. Enhanced Oil recovery in shale reservoirs by gas injection. Journal of Natural Gas Science and Engineering 22(2015): 8.
Sheng, J. J. and K. Chen, 2014. Evaluation of the EOR potential of gas and water injection in shale oil reservoirs. Journal of Unconventional Oil and Gas Resources 5: 1-9.
Theloy, C., S. A. Sonnenberg and J. B. Curtis, 2011. Contribution of natural fractures to production, Sanish field, north Dakota
Vincent, M., 2011. Restimulation of unconventional reservoirs: When are Refracs beneficial? Journal of Canadian Petroleum Technology 50(05): 36-52.
Wan, T., 2013. Evaluation of the EOR Potential in Shale Oil Reservoirs by Cyclic Gas Injection. MSC.
Wan, T., X. Meng, J. J. Sheng and M. Watson, 2014. Compositional Modeling of EOR Process in Stimulated Shale Oil Reservoirs by Cyclic Gas Injection, SPE-169069-MS.
Wan, T., J. J. Sheng and M. Y. Soliman, 2013. Evaluate EOR Potential in Fractured Shale Oil Reservoirs by Cyclic Gas Injection, SPE-168880-MS.
Wan, T., J. J. Sheng and M. Y. Soliman, 2013. Evaluation of the EOR Potential in Shale Oil Reservoirs by Cyclic Gas Injection, SPWLA-2013-MM.
Warren, J. E. and P. J. Root, 1963. The behavior of naturally fractured reservoirs. Society of Petroleum Engineers Journal 3(03): 245-255.
Wu, Y.-S., Y. Di, Z. Kang and P. Fakcharoenphol, 2011. A multiple-continuum model for simulating single-phase and multiphase flow in naturally fractured vuggy reservoirs. Journal of Petroleum Science and Engineering 78(1): 13-22.
Zangeneh, N., E. Eberhardt and R. M. Bustin, 2014. Investigation of the influence of natural fractures and in situ stress on hydraulic fracture propagation using a distinct-element approach. Canadian Geotechnical Journal 52(7): 926-946.

Zangeneh, N., E. Eberhardt and R. M. Bustin, 2015. Investigation of the influence of natural fractures and in situ stress on hydraulic fracture propagation using a distinct-element approach. Canadian Geotechnical Journal 52(7): 926-946.

537019 Enhancing Hydraulic Fracturing Productivity Via Model-Based Feedback Control

Abstract:

In hydraulic fracturing, the proppant-filled fracture length at the end of pumping strongly influences the fluid conductivity of natural oil and gas [1]. Therefore, it is very important to regulate proppant bank height and suspended proppant concentration across the fracture to increase the recovery of shale hydrocarbon [2, 3]. Currently, pumping schedule is designed offline and applied to a hydraulic fracturing process in an open-loop manner, which may lead to poor process performance if there are large disturbances and plant-model mismatch.

Motivated by these considerations, we focus on the development of a model predictive control framework for the design of pumping schedule to regulate the spatial variation of proppant concentration across the fracture at the end of pumping for conventional and unconventional reservoirs. In conventional oil reservoirs, high-viscosity fracturing fluids ensure that most of the proppant remains in suspension during the treatment and the closure process. Thus, it is sufficient to regulate the suspended proppant concentration along the fracture at the end of pumping. In unconventional reservoirs, however, predominantly low-viscosity (â€œslick-waterâ€) fluids are used and the proppant settles quickly forming a proppant bank, which will continue to grow until it reaches the equilibrium height; a state when the rate of proppant washout on top of proppant banks due to the shear force is equal to the rate of bank formation via proppant settling. To this end, we initially focus on the development of a first-principle model of a hydraulic fracturing process to obtain fundamental understanding of the proppant bank formation mechanism and its relationship to manipulated input variables such as proppant concentration and flow rate of the injected fracturing fluids. The optimal fracture geometry is obtained by applying Unified Fracture Design for conventional reservoirs [6] and a section-based optimization method for unconventional reservoirs [7]. Next, high-fidelity simulation data has been used to construct a data-based linear approximate model to design a Kalman filter that estimates unmeasurable states. Lastly, a model-based feedback controller is developed to achieve the uniform proppant bank height and suspended proppant concentration along the fracture at the end of pumping for conventional and unconventional reservoirs by explicitly taking into account the desired fracture geometry, type of the fracturing fluid injected, total amount of injected proppant, actuator limitations, and safety considerations.

References:

[1] Yang, S., Siddhamshetty, P., Kwon, J.S. (2017), Optimal pumping schedule design to achieve a uniform proppant concentration level in hydraulic fracturing. *Compt. Chem. Eng.*, 101, 138-147.

[2] Siddhamshetty, P., Yang, S., & Kwon, J. S. I. (2017). Modeling of hydraulic fracturing and designing of online pumping schedules to achieve uniform proppant concentration in conventional oil reservoirs. *Computers & Chemical Engineering. URL: https://doi.org/10.1016/j.compchemeng.2017.10.032.*

[3] Siddhamshetty, P., Kwon, J. S. I., Liu, S., & Valkó, P. P. (2017). Feedback control of proppant bank heights during hydraulic fracturing for enhanced productivity in shale formations. *AIChE Journal*, 64(05), 1638-1650.

[4] Gu, H., & Desroches, J. (2003, January). New pump schedule generator for hydraulic fracturing treatment design. *SPE Latin American and Caribbean Petroleum Engineering Conference*, Port-of-Spain, Trinidad and Tobago.

[5] Nolte, K. G. (1986). Determination of proppant and fluid schedules from fracturing-pressure decline. SPE Production Engineering, 1(04), 255-265.

[6] Economides, M.J., Oligney, R.E., Valko, P., 2002. Unified fracture design. Orsa Press.

[7] Liu, S., & ValkÓ, P. P. (2017). Optimization of spacing and penetration ratio for infinite conductivity fractures in unconventional reservoirs- A sectional based approach. *Society of petroleum engineers* (SPE-186107-PA).

523596 A Reduced Order Model for Optimizing Hydraulic Fracture Stimulation of Horizontal Wells

Abstract:

One approach to optimizing hydraulic fracturing stimulations of horizontal oil and gas wells is to maximize created fracture surface area and/or maximize the uniformity of stimulation across each fracturing stage. Here we present a rapidly-computing model to capture the impact of fluid flow and fluid leakoff on surface area and/or uniformity optimization while enabling optimization requiring hundreds or thousands of model evaluations. The key innovation for the rapidly-computing model is a novel concept we call the â€œcomposite viscosityâ€, which is essentially to lump all dissipative processes into an equivalent viscosity that allows the simulator to estimate the length, width and pressure based on a reduced order model that globally conserves mass, momentum, and energy. The results of these simulations show a tradeoff. On one hand, low viscosity leads to larger fracture surface area because high viscosity leads to larger fracture widths and the potential for more suppression of fractures by the stresses induced by their neighbors. However, low viscosity also leads to decreased fracture surface area due to higher fluid losses to the formation. This tradeoff leads to the existence of a fluid viscosity, specific to the properties of a given reservoir rock, that optimizes fracture surface area and/or uniformity.

530244 Prototype LIBS Sensor for Sub-Surface Water Quality Monitoring

Abstract:

â€œPrototype LIBS sensor for sub-surface water quality monitoring with applications in carbon storageâ€
D.A. Hartzler[1,2], J.C. Jain[1,2], D.L. McIntyre[1]

[1]*National Energy Technology Laboratory, U.S. Department of Energy, Pittsburgh, PA 15236, USA.*
[2]*AECOM, National Energy Technology Laboratory, Pittsburgh, PA 15236, USA.*

Geologic carbon storage (GCS) involves injection of CO2 into rock formations often containing high salinity brines. In the event of CO2 leakage, freshwater above the storage sites can potentially be contaminated with metals either originating from the brine or released by leaching of the formation rocks. An increase in the concentration of these proxy metals would warn of leakage well before the CO2 has reached the surface. Therefore, a Laser Induced Breakdown Spectroscopy (LIBS) sensor is being developed for sub-surface water quality monitoring. The sensor head is built using a low cost passively Q-switched (PQSW) laser, which is fiber coupled to a pump laser and a gated spectrometer. The ability to separate the LIBS excitation laser from the pump and spectrometer would allow the expensive and fragile components to remain at the surface while only the sensor head needs to be exposed to the hostile downhole environment. Spectrometer gating is achieved by triggering from a backwards traveling, nanosecond pulse coming to the surface through the pump fiber, eliminating the need for any electronic components to be located downhole. With the relative low cost and simplicity of the (laser) head, the sensor could be left in place for continuous long-term monitoring of a GCS site without the need to retrieve samples or equipment from downhole. A prototype has been constructed using off the shelf components along with a custom, monolithic PSQW laser. Laboratory testing of the prototype shows Limits of Detection (LODs) comparable to those of actively Q-switched systems and has allowed verification that the fiber coupling and triggering schemes perform as desired. Work to produce a miniaturized submersible version is in progress.

533566 Catalytic Hydrogenation of Furfural over Rumop: Probing Bimetallic andÂ Compositional Effects on Catalyst Performance

Abstract:

Catalytic Hydrogenation of Furfural over RuMoP: Probing Bimetallic and Compositional Effects on Catalyst Performance

Yolanda Bonita, Timothy P. Oâ€™Connell, and Jason C. Hicks*

Department of Chemical and Biomolecular Engineering, University of Notre Dame, 182 Fitzpatrick Hall, Notre Dame, IN 46556

The synthesis of furfuryl alcohol from furfural uses an environmentally toxic copper chromite catalyst in the process. As possible alternatives, Cu-based catalysts such as supported Cu, bimetallic Cu-Ni, Cu-Co, Cu-Fe, and Cu-Ru have been studied to replace copper chromite. However, none of these catalysts surpass the high selectivity exhibited by copper chromite for furfural hydrogenation to furfuryl alcohol.

We have recently discovered a variety of novel transition metal phosphides that are highly selective to furfuryl alcohol and active at near ambient temperature. Contrary to the literature that suggested group VIII metals are not selective to furfuryl alcohol due to the binding configuration, we have synthesized group VIII metal (i.e. Ru) phosphides that exhibit high selectivity to furfuryl alcohol, similar to that of copper chromite and higher than other Cu-based catalysts. Moreover, when comparing pseudo-first order rate constants, we see an increase of > 5 times in the site-normalized, initial rate and 99% selectivity for bimetallic RuMoP. This phenomena was also observed in our previous studies (i.e. hydrodeoxygenation reaction)[1-4], which showcases bimetallic character and potential control of the catalytic properties based on catalyst composition. Indeed, we have observed direct influence on the activity and selectivity by varying the composition of $Ru_xMo_{2-x}P$, where the measured pseudo-first order rate constant was enhanced for Ru-rich materials (x = 1.2). In this presentation, we will present a thorough examination of bimetallic RuMoP as a furfural hydrogenation catalyst through detailed characterization and kinetic experiments.

References

1. Rensel, D. J., Rouvimov S., Gin M. E., Hicks J. C. Highly selective bimetallic FeMoP catalyst for C-O bond cleavage of aryl ethers. *J. Catal.* Sep 2013;305:256-263.
2. Rensel, D. J., Kim J., Bonita Y., Hicks J. C. Investigating the multifunctional nature of bimetallic FeMoP catalysts using dehydration and hydrogenolysis reactions. *Appl. Catal. A.* Aug 25 2016;524:85-93.
3. Rensel, D. J., Kim J., Jain V., Bonita Y., Rai N., Hicks J. C. Composition-directed $Fe_xMo_{2-x}P$ bimetallic catalysts for hydrodeoxygenation reactions. *Catal. Sci. Technol.* 2017;7:1857-1867.
4. Bonita, Y., Hicks J. C. Periodic trends from metal substitution in bimetallic Mo-based phosphides for hydrodeoxygenation and hydrogenation reactions. *J. Phys. Chem. C.* 2018, ASAP.

532119 Etherification of 5-Hydroxymethylfurfural Using Zeolite Catalysts

Abstract:

Biomass is a promising feedstock for the production of modern chemicals. In this regard, polymers represent an attractive class of potential biobased materials, and it may be possible to synthesize a variety of polymeric materials starting from 5-hydroxymethylfurfural (HMF). In this study, we show that H-BEA zeolite is highly active and selective for the production of a variety of HMF ethers which would be suitable for subsequent polymerization.

Several catalysts were evaluated for their etherification activity, including: H-BEA and H-ZSM-5 zeolites, Amberlyst-15 (A-15), and Î³-Al_2O_3. These catalysts were chosen to evaluate a variety of active sites and pore structures. Catalysts possessing BrÃ¸nsted acid sites (*i.e.*, H-ZSM-5 and H-BEA zeolites, and A-15) showed the highest selectivities to the ether product, with H-BEA achieving nearly quantitative selectivity. We hypothesize that the improved selectivity observed for H-BEA, compared with A-15, is due to confinement effects that inhibit both levulinate formation and biomolecular condensation (*i.e.*, resinification) reactions.

The rate of etherification is nearly constant for the series of ethanol, *n*-butanol, cyclohexanol, and phenol. Moreover, while the etherification rate saturates at high HMF concentrations, there is little variation in the rate across for alcohol concentrations ranging from 0.1 to 10 mol L^{-1}. These kinetics data, coupled with the invariance of the rate with varying alcohol data, suggest that the reaction proceeds through an oxocarbenium ion transition state that does not involve the etherifying alcohol.

532366 Mechanistic Insights into the Hydrogenolysis of Levoglucosanol over Bifunctional Platinum Silica-Alumina Catalysts in Tetrahydrofuran Solvent

Abstract:

The cellulose-derived platform molecule levoglucosenone (LGO) can be converted to a variety of high-value oxygenated chemicals. [1] LGO can be quantitatively hydrogenated to levoglucosanol (Lgol) over metal catalysts. [2] Herein, we report on the hydrogenolysis of Lgol to tetrahydrofurandimethanol (THFDM) and tetrahydropyran-2-methanol-5-hydroxyl (THP2M5H) over bifunctional platinum or palladium catalysts supported on silica-alumina in tetrahydrofuran solvent. [3,4] THFDM is an α,ω-diol and potential polymer precursor, while both THFDM and THP2M5H can be further upgraded to the commodity chemical 1,6-hexanediol.

A combination of experimental reaction kinetics studies and first principles calculations were used to investigate the mechanism and the roles of the metal and acid sites in this reaction. THFDM and THP2M5H are produced in parallel pathways, and the ketone intermediate THP2M5one is a precursor to THP2M5H. [3] The THFDM cis/trans ratio is independent of the Lgol threo/erythro ratio, suggesting that the mechanism passes through an acyclic intermediate which erases the stereochemistry of the Lgol reactant. ^{13}C radio-labeling was used to confirm the ring rearrangement forming THFDM. The catalyst undergoes moderate, reversible deactivation with time-on-stream in a continuous flow reactor. The reaction rates and product selectivities are comparable at 1.1% and 5.3% Pt loadings, indicating that the reaction rate is acid-limited at these metal loadings; the metal-catalyzed hydrogenation step is not rate-limiting (Figure 1). [4] The measured zero-order dependence in hydrogen indicates that the rate-limiting acid-catalyzed step precedes the fast hydrogenation step. As the Pt loading in Pt/SiAl catalysts is decreased, or when the bare SiAl support is separated from a Pt/SiO_2 catalyst in a dual-layer configuration, the selectivity towards identified products decreases (Figure 1). These results suggest that degradation reactions are avoided when the reactive intermediates formed over acid sites are rapidly hydrogenated over metal sites. First principles simulations were performed to investigate the energetics of the proposed reaction pathway consistent with the experimental results.

Figure 1. *Lgol consumption rate (left y-axis) and product selectivity (right y-axis) versus Pt loading for Pt/SiO_2-Al_2O_3 catalysts.*

These insights into the mechanism of Lgol hydrogenolysis provide a foundation for rationally designed processes to produce high-value chemicals from lignocellulosic biomass using bifunctional metal-acid catalysts.

References.

1. Cao, F.; Schwartz, T. J.; McClelland, D. J.; Krishna, S. H.; Dumesic, J. A.; Huber, G. W., Dehydration of cellulose to levoglucosenone using polar aprotic solvents. *Energy & Environmental Science* **2015,** *8* (6), 1808-1815.

2. He, J.; Huang, K.; Barnett, K. J.; Krishna, S. H.; Alonso, D. M.; Brentzel, Z. J.; Burt, S. P.; Walker, T.; Banholzer, W. F.; Maravelias, C. T.; Hermans, I.; Dumesic, J. A.; Huber, G. W., New catalytic strategies for [small alpha],[small omega]-diols production from lignocellulosic biomass. *Faraday Discussions* **2017,** *202* (0), 247-267.

3. Krishna, S. H.; McClelland, D. J.; Rashke, Q. A.; Dumesic, J. A.; Huber, G. W., Hydrogenation of levoglucosenone to renewable chemicals. *Green Chemistry* **2017,** *19* (5), 1278-1285.

4. Krishna, S. H.; Assary, R. S.; Rashke, Q. A.; Schmidt, Z. R.; Curtiss, L. A.; Dumesic, J. A.; Huber, G. W., Mechanistic Insights into the Hydrogenolysis of Levoglucosanol over Bifunctional Platinum Silica-Alumina Catalysts. *ACS Catalysis* **2018,** 3743-3753.

538548 Hydrothermal Stability of Chloromethyl Polystyrene Based Solid Acid Catalysts and Mechanism of Cellulose Hydrolysis

Abstract:

Cellulose is one of the major components of lignocellulosic biomass and the most abundant renewable polymer on earth. As such it could serve as a sustainable feedstock for the production of commercially important fuels and chemicals. However, cellulose conversion to smaller platform molecule is the major bottleneck for economically competitive utilization of lignocellulosic biomass.

The most common processes to depolymerize cellulose employ liquid acid or enzyme catalyzed hydrolysis to produce glucose. However, issues such as reactor corrosion, slow reaction rates, and inability to recycle limits the cost-effective conversion of cellulose. Alternatively, solid acids have been proposed as reusable catalysts for cellulose hydrolysis. Introduction of binding groups in the catalyst structure is hypothesized to increase the affinity between the solid acid and cellulose, brining the glycosidic bonds within contact with catalytic sites resulting in a higher hydrolysis rate. This hypothesis finds support in adsorption studies of small cellulosic molecules such as cellobiose and glucose. However, the mechanism of such catalytic activity has not been elucidated.

Chloromethyl polystyrene based bifunctional catalysts have showed high activity towards cellulose depolymerization. It is hypothesized the chlorine participates in hydrogen bonding with the hydroxyl groups in cellulose chains. However, the chloromethyl groups is rather label and could hydrolyze in hydrothermal conditions generating hydrochloric acid. Hydrochloric acid could also catalyze cellulose depolymerization, but its effect on the overall catalytic mechanism has not be evaluated.

In this work we studied partially sulfonated chloromethyl polystyrene catalyst for cellulose hydrolysis. Structural analysis of catalyst revealed that synthesis procedure results in little to no chlorine available on the external surface, but presence of hydroxyl groups, which can also participate in hydrogen bonding with cellulose. The catalyst showed high activity towards hydrolyzing cellulose resulting in levulinic acid and formic acid as major products. However, analysis of the solid and the liquid media revealed chloromethyl group hydrolysis and hydrochloric acid generation. This observation raised the question if the catalyst is indeed acting as a solid acid.

In order to elucidate the mechanism of hydrolysis we subjected the catalyst to hydrothermal reaction conditions in the absence of cellulose. Monitoring pH and chloride ion concentration, we developed a kinetic profile of hydrochloric acid generation. Cellulose hydrolysis kinetic models involve ion concentration as a preexponential factor on the expression for the reaction rate constant. We hypothesized that incorporating the time dependence of the acid concentration in the kinetic model of cellulose hydrolysis to glucose and subsequent dehydration to HMF, levulinic acid, and formic acid would be able to accurately predict experimental results. An acceptable agreement between model and experimental product yields pointed to the liquid acid acting as the primary catalyst, obscuring any catalytic activity, if any, of the solid catalyst. Further support was obtained by carrying out hydrolysis by a titrated solid acid catalyst, in which the sulfonic acid groups have been ion exchanged and, thus, rendered inactive; similar yields were obtained as the non-titrated catalyst. The results in this investigation highlight the need for systematic study of proposed solid acids for cellulose hydrolysis and consideration of potential release of homogeneous acids in the reaction media.

538054 Glycerol Transfer-Hydrogenation of Levulinic Acid Using Ru and Ir Carbene Organometallics Immobilized on Active Hydrotalcites

Abstract:

The mechanism and kinetics of levulinic acid (LA) hydrogenation to gamma-valerolactone (GVL) have been extensively studied over the last decade. Despite improvements in selectivity and the use of alternative hydrogen sources such as formic acid and secondary alcohols, economic production of GVL remains hindered by slow reaction rates and the high cost of current hydrogen sources. The use of glycerol, a waste byproduct in biodiesel synthesis, as a hydrogen donor could alleviate the economic challenges associated with GVL production. Herein we tested hydrogenation/deoxygenation of LA under dilute aqueous to neat conditions using Ru and Ir N-heterocyclic carbene (NHC) organometallic catalysts while also varying pH with potassium hydroxide. A total of 7 Ru and Ir homogeneous catalysts were screened for activity, the most active catalysts were then immobilized onto tunable hydrotalcites (HT). A 3:1 ratio of Mg:Al HT was used as a basis for all HT catalytic reactions while substitutions of Al with Fe, Ni, and Co were tested for improved catalytic activity. Products were identified using LCMS, HPLC, NMR, GCMS, and FTIR. The products formed and reaction rates were pH-dependent; acidic conditions favored ketal formation between LA and glycerol whereas alkaline conditions with an Ir(I) complex favored GVL and lactic acid formation at a turnover frequency (TOF) above 66,000 h^{-1} at 150ËšC for GVL. The Ir(I) catalyst achieved unprecedented reaction rates with glycerol, therefore, a comparison to reaction rates achieved with formic acid was demonstrated and showed negligible differences in rates and selectivity. Upon immobilizing the NHC catalysts on HT, the non-active Ru and Ir catalysts achieved activity over 10,000 h^{-1}. Under the same conditions the immobilized Ir(I)HT rate was reduced to 15,000 h^{-1} for GVL. The increase in activity from the other Ru and Ir(III) complexes tested are believed to be a result of the HT Lewis acid sites.

530511 Paired Electrocatalytic Hydrogenation and Oxidation of 5-Hydroxymethylfurfural for Efficient Production of Biorenewable Monomers

Abstract:

Electrochemical synthesis is a promising route for sustainable chemical production; however, its widespread application is hindered by economic limitations incurred from high capital and operating costs. In this presentation, we report highly efficient electrocatalytic conversion of 5-(hydroxymethyl)furfural (HMF) to biobased monomers by pairing HMF reduction and oxidation half reactions in a single electrochemical cell. HMF hydrogenation to 2,5-bis(hydroxymethyl)furan (BHMF) was achieved under mild electrolyte conditions and ambient temperature using a Ag/C cathode. Meanwhile, HMF oxidation to 2,5-furandicarboxylic acid (FDCA) with ~100% efficiency was facilitated under the same conditions by a homogeneous nitroxyl radical redox mediator, together with an inexpensive carbon felt anode. The selectivity and efficiency for Ag-catalyzed BHMF formation were sensitive to cathode potential, owing to competition from HMF hydrodimerization reactions and water reduction (hydrogen evolution). Moreover, the carbon support of Ag/C was active for HMF reduction and contributed to undesired dimer/oligomer formation at strongly cathodic potentials. As a result, high BHMF selectivity and efficiency were only achieved within a narrow potential range near â€“1.3 V. Fortunately, the selectivity of redox-mediated HMF oxidation was insensitive to anode potential, thus allowing HMF hydrogenation and oxidation half reactions to be performed together in a single cathode-potential-controlled cell. Electrocatalytic HMF conversion in a paired cell achieved high molar yields of BHMF and FDCA, and nearly doubled electron efficiency compared to the unpaired cells. Future research will be focused on rational design of an electrochemical flow reactor, in which the use of a membrane electrode assembly (MEA) can significantly lower the internal resistance and reduce the total energy demand.

525127 A Full FurfuralÂ Utilization over Ni/SiO_2 Catalysts

Abstract:

Furfural can be mass-produced from lignocellulose biomass and can be a platform chemical for producing valuable chemicals. In this study, we examine Ni/SiO_2 catalysts for the conversion of furfural under a hydrogen atmosphere. The reaction products include furfuryl alcohol (FA), furan, and propylene as major; butanal and other hydrogenated products, indicating the reaction routes involving side chain hydrogenation, decarbonylation, and ring opening. The reactivity and the product selectivity depend on the reaction temperature and the Ni particle size. The pretreatment by including calcination prior to hydrogen reduction leads to Ni/SiO_2â€"CR with large Ni particles (~ 15 nm) and a high selectivity to FA at below 200 °C. The Ni/SiO_2â€"R is hydrogen-pretreated without the calcination and it contains small Ni particles (~ 5 nm) and exhibits relatively high selectivity to furan. Moreover, the furan can be reacted seriously to produce propylene and CO by this Ni catalyst. The turnover frequency (TOF) of furfural conversion is 239 and 408 h^{-1}, respectively, on Ni/SiO_2-CR and Ni/SiO_2-R at 175 °C, when the former shows 100% selectivity to FA and the latter exhibits a selectivity of around 38% and 62%, to FA and furan, respectively. The results indicate that the furfural hydrotreating reaction over Ni catalysts is with structure sensitive and give a full furfural utilization, either to FA (a biofuel), or to propylene and CO (a chemical feedstock).

527041 Two Stage Stochastic Programming Modeling Approach for Energy Storage System Operation Under Uncertainty

Abstract:

Employing renewable energy sources has become an irreversible trend in the current energy systems of various sectors including process industries. Their decision-makers should cope with often occurring mismatches between varying energy supply and demand over multiple time periods. Considering the inherent variability of renewable output, it would be often witnessed that surplus energies from renewable sources is simply thrown away without being used. It is thus understandable to see the introduction of energy storage system (ESS) for the sake of the economic benefit of using the surplus again later. We can increase the penetration of renewable energies by installing the ESS. On the other hand, it is also right to point out that there is much to be explored in maximizing the potential of ESS with other entities in the energy system. In order to transform the expectation into reality, the corresponding decision-making framework should be rigorously designed. As an alternative, this paper investigates the impact of variation on the design and operation of ESS. Specifically, the design of ESS under uncertainty is transformed into a two-stage stochastic programming model. The stochastic model was formulated in terms of a mixed integer linear programming (MILP) problem. As illustrated in numerical examples, the proposed modeling approach allows us to gain the insight on renewable energy systems. More similar works are expected to establish the robust energy system in process industries.

527226 Theoretical and Experimental Investigation of the Microbial Degradation of Solitary Oil Microdroplets

Abstract:

Natural seeps and accidental releases of crude oil in the sea result in clouds of droplets that are carried away by underwater sea currents. The droplets may be created either at the sea surface during the breakup of an oil slick by sea waves, or at the seafloor during the atomization of live crude oil extruding at sufficiently high speed from a natural crack or a broken wellhead. Drifting droplet clouds disturb the established ecosystem dynamics and pose a high risk of toxic effects to many marine species. It is therefore of crucial importance to understand and quantify the factors that determine the retention of the dispersed droplets within the seawater column. Dissolution, biodegradation and aggregation with marine snow are the main physical processes that rule the fate of dispersed oil droplets in marine waters.

With regard to the biodegradation process, microbes have developed three fundamental strategies for accessing and assimilating oily substrates. Depending on their affinity for the oily phase and ability to proliferate in multicellular structures, microbes might either attach to the oil surface and directly uptake compounds from the oily phase, or grow suspended in the aqueous phase consuming solubilized oil, or form three-dimensional biofilms over the oil-water interface (Kapellos, 2017).

We have recently developed a compound particle model for the biodegradation of solitary oil microdroplets moving through a water column (Kapellos et al., 2018). The compound particle is of the core-shell type and consists of an oily core that is successively surrounded by a bioreactive skin of negligible thickness and another bioreactive shell of finite thickness. The bioreactive skin represents a thin layer of microbes that uptake oil directly from the oily phase, whereas the bioreactive shell represents a distinct biofilm phase. The new model accounts for all three modes of biodegradation: direct interfacial uptake, bioreaction in the bulk aqueous phase, and bioreaction in a biofilm formed around the droplet. A closed-form correlation has been established for the overall dissolution rate for a given particle configuration and two coupled ordinary differential equations have been derived for the evolution of the dimensions of the compound particle. Numerical analysis is currently used to extend the domain of validity of the model by relaxing key hypotheses. In this presentation, major findings of the theoretical analysis will be discussed in detail and in conjunction with preliminary results from the biodegradation of hexadecane droplets by Marinobacter sp. in a microfluidic setting.

Acknowledgements: This work has received funding from the European Unionâ€™s Horizon 2020 research and innovation programme under the Marie Sklodowska-Curie grant agreement No 741799 (project "OILY MICROCOSM").

References:

Kapellos, G.E. "Microbial strategies for oil biodegradation", In *Modeling of Microscale Transport in Biological Processes*; Becker, S.M., Ed.; Academic Press, pp. 19-39 (2017).

Kapellos, G.E., Paraskeva, C.A., Kalogerakis, N. and Doyle, P.S. "Theoretical insight into the biodegradation of solitary oil microdroplets moving through a water column", *Bioengineering*, 5:15, (2018).

530591 Modeling Faradaic Capacitive Deionization with Redox Active Porous Electrodes

Abstract:

Capacitive deionization (CDI) is an attractive technology for brackish water treatment, due to the low energy consumption at low source water salinity level compared to conventional water treatment technologies such as reverse osmosis or distillation.

It has been shown that the surface charge on the porous carbon electrodes has significant impact on CDI performance. By introducing immobilized chemical groups, the point of zero charge of carbon materials can be tuned, thus enhanced or inversed CDI performance could be obtained. Redox active electrodes have been investigated in CDI applications recently. The Faradaically induced chemical charges add to electronic charges within the electric double layers formed at the electrode surfaces to provide a much higher adsorption capacity than attained in purely capacitive processes, which is denoted as Faradaic capacitive deionization (FaCDI) here. The redox active ligand can be considered as variable chemical charges that depend on the cell voltage, as opposed to the fixed charges introduced by chemical modification.

Here, we developed a model which incorporates the effect of variable chemical charge due to general faradaic reactions of redox species immobilized on the electrode surfaces. The theory describes the variable chemical charge by the generalized Frumkin-Butler-Volmer equation across the stern layer as an extension of current electrosorption theory. As a first approximation, the model assumes the spacer channel is well-mixed with uniform electrosorption in each electrode. The model is able to predict the experimentally observed enhanced and inverted performance of CDI cells. Furthermore, the deionization performance of FaCDI cells is shown to be superior to that of CDI and enhanced-CDI systems with equilibrium adsorption capacities 50-100% higher than attained with CDI systems. By properly choosing the equilibrium potentials of the redox active material, FaCDI is able to achieve the same ion adsorption capacity at smaller voltages compared to conventional CDI, which indicates the promise of obtaining higher salt adsorption capacity with lower energy input through the use of FaCDI materials.

The model enables us to optimize the operation parameters of the process, such as flow rate, applied voltage as well as the design of CDI cells, such as the surface to volume ratio of the device, in order to obtain the optimal deionization performance.

532901 Finding the Ideal Feedstock for Biodiesel Production Via Generalized Kinetic Model for Transesterification and Saponification

Abstract:

The hunt for an assortment of energy sources is just about as old as humanity's need for energy. This search has been escalated by an expanding industrialization since the 19th century. The internal combustion engine is a prime example. Even though fuels obtained from fossil sources have been a prime choice since the onset of these engines, biofuels especially biodiesel have also found some interest. This interest has particularly gained momentum in the past few decades due to rapidly increasing energy demand and depletion in fossil fuel resources.

Biodiesel is a non-toxic, biodegradable, and environmentally benign fuel that can be used in diesel engines. The most commonly used method for the production of biodiesel is transesterification. It involves the reaction of oil with an alcohol in the presence of a catalyst to form fatty acid alkyl ester (biodiesel) and glycerol. Although, biodiesel is a promising alternative fuel, its production poses significant challenges due to its higher costs compared to that of petroleum-based diesel. Typically one of the major contributors to this issue is the cost of the feedstock, which accounts for approximately 60 to 80% of the total production cost.

Even though the literature is replete with the studies about modelling and optimization of transesterification [1, 2, 3, 4], very few researchers have focused their attention on the selection of feedstock for biodiesel production. Although, a few comparative studies have been reported in the literature [2, 5], all of them focus on static optimization, thereby neglecting the kinetics of biodiesel production. Therefore, in order to address these issues, we aim to achieve the following objectives:

1. Find optimal operating conditions of the reactor for biodiesel production using different vegetable oils as feedstock.
2. Develop meta-models depicting the effect of the different vegetable oils/their blends on the properties of the biodiesel produced using these feedstocks.
3. Use the developed meta-models to find the best possible feedstock for a particular country.

The purpose of this study is to find optimal feedstock (vegetable oil blend) for a particular country based on the standard property specifications and the availability of oils in that country. To this end, we use the generalized kinetic model, developed in our previous work [6] to find optimal conditions with 4,500 different blends of six major oils *viz.*, Jatropha, Linseed, Olive, Palm, Rapeseed, and Sunflower oil as feedstock. We employ a property based objective function (considering density, viscosity, cold point, pour point, cetane number, higher heating value, lower calorific value, and cold filter plugging point of biodiesel) to solve these optimization problems. We then employ correlations from the literature [3] to calculate properties of biodiesel formed using a particular feedstock and at corresponding optimal conditions. This is followed by developing meta-models with the computed properties of biodiesel as output variables and different characteristics of the feedstocks as the input variables. We use five different approaches to characterize the feedstock *viz.* a) average chain length and the unsaturation degree of the fatty acids that constitute the oil (adopted from [7], number of dimensions (N) = 2); b) average chain length, weighted unsaturation degree calculated based on the triglyceride composition of oil (proposed in this work, N = 2); c) triglyceride composition (proposed in this work, N = 34); d) weights on the considered oils for the different blends (proposed in this work, N = 6); and e) properties of oils, *i.e.* viscosity, density, heat capacity, enthalpy of vaporization, and vapour pressure calculated based on the triglyceride composition of oil (proposed in this work, N = 5). Finally, we use the best performing meta-models to obtain optimal feedstock for production of biodiesel in a particular country.

The predictions made using the developed meta-models are in accord with those made using the original models. We quantify this agreement, by computing the Percentage Average Absolute Error (PAAE) as follows,

PAAE = (Average Absolute Error/Average Response)x100

The error values are below 1% for most of the cases with the maximum value at 2.55%, which shows excellent prediction performance. Moreover, we use a normalized error metric to compare the performance of the meta-models developed using different data sets in this study. We observed that the meta-models developed using triglyceride composition and weight distribution of oils in the blend as input variables performed superior to the other meta-models. However, owing to the large number of dimensions in the former case, it is apt to use the latter. Finally, we used the meta-models to obtain optimal feedstock blend for a particular country. We analysed different scenarios based on the standard specification imposed on the biodiesel quality, availability and type of feedstock (*i.e.* edible or non-edible). We observed that the quality of biodiesel produced can change significantly based on the considered scenarios, thereby highlighting the importance of such a study. Besides, we also observed that it is possible to produce premium quality biodiesel even with just non-edible oils. Such a result can definitely bolster the efforts for commercialization of biodiesel production as the use of non-edible oils can significantly reduce the costs associated with the biodiesel production process.

REFERENCES

[1] V. Singh, K. Solanki, M. N. Gupta, Process optimization for biodiesel production, Recent patents on biotechnology 2 (2) (2008) 130{143.

[2] J.Yang, T. Astatkie, Q. S. He, A comparative study on the effect of unsaturation degree of camelina and canola oils on the optimization of bio-diesel production, Energy Reports 2 (2016) 211-217.

[3] S. Pinzi, D. Leiva, G. Arzamendi, L. Gandia, M. Dorado, Multiple response optimization of vegetable oils fatty acid composition to improve biodiesel physical properties, Bioresource technology 102 (15) (2011) 7280-7288.

[4] R. S. Kumar, K. Sureshkumar, R. Velraj, Optimization of biodiesel production

from manilkara zapota (l.) seed oil using taguchi method, Fuel 140 (2015) 90-96.

[5] T. Issariyakul, A. K. Dalai, Comparative kinetics of transesterication for biodiesel production from palm oil and mustard oil, The Canadian Journal of Chemical Engineering 90 (2) (2012) 342-350.

[6] P. Chhabra, S. Mosbach, M. Kraft, I. A. Karimi, A generalized kinetic model for transesterication and saponification (2017).

URL https://aiche.confex.com/aiche/2017/meetingapp.cgi/Paper/500597

[7] E. G. Giakoumis, A statistical investigation of biodiesel physical and chemical properties, and their correlation with the degree of unsaturation, Renewable Energy 50 (2013) 858-878.

530255 Accelerating the Generation of Coal Power Plant Property Models

Abstract:

Power plants are complex systems that are represented by first principles or empirical models. As power plants are developed with new technologies outside of well-studied thermodynamic regions, data-driven approaches have been used to model these complex systems [5]. For example, thermophysical properties or kinetic reaction models can be explained to an arbitrary level of complexity with an empirical model. The challenge is that these process models are not well understood, and it is not always intuitive what functional forms can be utilized to represent these systems. When optimizing proposed model systems, the efficiency of power plants relies on having accurate models. To maximize efficiency as well as to maintain safe operating conditions, it is essential that complex kinetic and thermophysical relationships are represented accurately, even though determining such a representation may be computationally expensive.

In this work we address the problem of speeding up the model generation process. In particular, we look at accelerating best subset selection algorithms with the use of GPU parallel computing. Best subset selection is a regression problem that seeks to determine a small set of features that best relates a set of inputs to an output [7]. There are many exact and heuristic approaches that can be used to solve the best subset selection problem, such as branch-and-bound [4], forward and backward selection [3], exhaustive search, the lasso [7], and mixed-integer optimization [6, 2, 1]. The proposed approach speeds up the model building process, allowing us to consider a larger feature space and generate more realistic models. The proposed methodology is compared against best subset selection techniques as well as the Automated Learning of Algebraic MOdels (ALAMO) methodology [2].

References

[1] D. Bertsimas, A. King, and R. Mazumder. Best subset selection via a modern optimization lens. The Annals of Statistics, 44:813–852, 2016.

[2] A. Cozad, N. V. Sahinidis, and D. C. Miller. Automatic learning of algebraic models for optimization. AIChE Journal, 60:2211–2227, 2014.

[3] M.A. Efroymson. Multiple regression analysis. Mathematical methods for digital computers, pages 191–203, 1960.

[4] C. Gatu and EJ Kontoghiorghes. Branch-and-bound algorithms for computing the best-subset regression models. Computational and Graphical Statistics, 15:139–156, 2006.

[5] S Lu and BW Hogg. Dynamic nonlinear modelling of power plant by physical principles and neural networks. International Journal of Electrical Power & Energy Systems, 22.

[6] A. Miller. Subset selection in regression. CRC Press, 2002.

[7] R. Tibshirani. Regression shrinkage and selection via the lasso. Royal Statistical Society, Methodological:267–288, 1996.

524382 Dynamic Optimisation of Water-Injection Wells Operation for Enhanced Oil Production from a Mature Oil and Gas Field

Abstract:

Field engineers in the oil and gas industry are constantly faced with the challenge of maintaining profitability amidst several operational constraints whose effects span a wide spectrum of timescales. Given the inevitably high environmental and financial stakes associated with the exploration and production of an oil and gas prospect, there is a strong incentive to enhance hydrocarbon recovery and production via systematic, model-based optimisation approaches [1-3]. In further response to this challenge, the application of sophisticated simulation methodologies to comprehensively ocapture the reservoir behaviour, multiphase flows (in wellbores and flowlines) [4-5] and gas-oil-water separation in the processing facilities is of exceptional industrial as well as academic interest. Liquid loading in gas wells and artificial lift design considerations, reservoir pressure maintenance via water injection, gas/water coning during production from deviated wells, pressure drop and liquid holdup of multiphase mixtures in highly deviated flowlines are some of the specific features of the problem which are frequently encountered and known to affect its complexity.

Simulation of these prevalent subsurface and surface phenomena does not always guarantee an accurate prediction of the onset of these problems, let alone the trouble-free operation of an existing petroleum field. In order to tackle the shortcomings and thus improve the fidelity of the currently employed models, it is necessary to also combine robust optimisation methods with these oil and gas CFD simulations [6-7]. This combination of simulation and optimisation algorithms increases the order of complexity due to model nonlinearity, non-convexity and the presence of discrete variables. There has been an increasing number of contributions in the application of optimisation techniques to field production, and this can be attributed to the advances in the development of specialised algorithms and accompanying computational power. To date, most of these attempts have focused on fields undergoing primary production through predominantly vertical wells [1-3].

This paper advances the state of the art in this field by formulating a production optimisation problem, via simultaneous consideration of production and injection wells with both vertical and deviated (inclined/horizontal) well geometries. A field undergoing secondary production (Enhanced Oil Recovery, EOR) is considered in this study. We use multiphase flow and reservoir simulation software in order to compute flowrates and pressure drops in wellbores and flowlines of the considered production network; extra measures are taken to ensure all flowlines are free from hydrate and wax at the prevalent temperature and pressure conditions. The problem is solved as a nonlinear program (NLP), comprising an economic objective function and several constraints to ensure operational feasibility. The adopted optimisation technique yields a reasonable increase in oil production compared to the usual and direct application of a black-box simulator. This also translates to an increased NPV with the current oil price, thus demonstrating the efficiency of the proposed method as a value addition tool.

REFERENCES

[1] Gunnerud, V., Conn, A. and Foss, B., 2013. Embedding structural information in simulation-based optimization. *Comput. Chem. Eng.* **53**: 35–43.
[2] Gerogiorgis, D.I., Georgiadis, M., Bowen, G., Pantelides, C.C. and Pistikopoulos, E.N., 2006. Dynamic oil and gas production optimization via explicit reservoir simulation. *Comput. Aided Chem. Eng.*, **21**: 179–184.
[3] Tavallali, M.S., Karimi, I.A., Teo, K.M., Baxendale, D. and Ayatollahi, S., 2013. Optimal producer well placement and production planning in an oil reservoir. *Comput. Chem. Eng.*, **55**: 109–125.
[4] Epelle, E.I. and Gerogiorgis, D.I., 2017. A multiparametric CFD analysis of multiphase annular flows for oil and gas drilling applications. *Comput. Chem. Eng.*, **106**: 645–661.
[5] Epelle, E.I. and Gerogiorgis, D.I., 2018. Transient and steady state analysis of drill cuttings transport phenomena under turbulent conditions. *Chem. Eng. Res. Des.* **131**: 520-544.
[6] Gunnerud, V. and Foss, B., 2010. Oil production optimization—A piecewise linear model, solved with two decomposition strategies. *Comput. Chem. Eng.*, **34**(11): 1803–1812.
[7] Kosmidis, V.D., Perkins, J.D. and Pistikopoulos, E.N., 2004. Optimization of well oil rate allocations in petroleum fields. *Ind. Eng. Chem. Res.* **43**(14): 3513–3527.

537394 Voluntary Oil Well Ignition As a Blowout Response: Analysis of Factors Influencing Viability

Abstract:

While most off-shore drilling use conventional steel rigs, for large reservoirs in shallow waters, artificial islands make an attractive alternative for offshore drilling because it offers more space, lower environmental impact, greater personnel safety, and lower cost. Drilling from an off-shore island may allow voluntary ignition as a blowout response if the oil contains sufficient gas. For voluntary ignition to be a viable blowout response, wellhead burning â€" that depends on a number of factors â€" needs to be nearly 100% efficient.

In this work, we investigate wellhead burning as a planned response to oilwell blowouts. The work presents a case study of proposed wellhead burning of 880 GOR, 27° API oil, should a blowout occur. The work takes into account some of the concerns regarding possible significant oil spill on Beaufort Sea as a result of incomplete combustion of the blowout effluent.

Burning of two-phase mixtures of oil and gas is a complex phenomenon that depends on the fraction of gas in the mixture, percentage of oil being entrained as liquid droplets in the gas stream, the size of the droplets, their residence time in the flame, among other parameters. Available models and correlations for the influence of these parameters on burning efficiency are examined in view of their relevance to wellhead burning.

This work will demonstrate that for some oil wells with high enough GOR, wellhead burning could be efficient enough for to be a reasonable response to a well blowout. In the studied case, a GOR of 880 combined with high total flowrate of the mixture, strongly implies nearly complete combustion of the well effluent during a blowout. The case study helped us formulate an algorithm to examine whether any particular well could be a possible candidate for wellhead burning as a planned response to a blowout. The approach suggests determining a wellâ€™s worst case discharge (WCD), estimating fluid velocities at well exit, estimating adiabatic flame temperature, and performing an energy balance to determine if all oil drops will evaporate.

536581 Integrated Forward Osmosis-Membrane Distillation Process for Sustainable Treatment of High Tds Produced Waters

Abstract:

Horizontal drilling and hydraulic fracturing make the extraction of tightly bound hydrocarbons from shale formations economically feasible. Brine management has become a key issue for the future of unconventional onshore oil and gas industry. In recent years, forward osmosis (FO) and membrane distillation (MD) have been widely investigated for treating wastewaters containing high levels of total suspended solids (TDS). Here we aim to demonstrate the feasibility of a hybrid FO-MD process for water recovery from high TDS produced waters, obtained from Southwestern Energy operations in the Marcellus shale. In FO-MD hybrid system, the FO process draws water from the feed solution while MD process reconcentrates the diluted draw solution and produces pure water. The integration of FO-MD processes combines the advantages of both processes and can provide high permeate quality and low membrane fouling. Cellulose triacetate (CTA) obtained from HTI (Albany, OR) and ethylene chlorotrifluoroethylene (ECTFE) obtained from 3M (Maplewood, MN) were employed as FO and MD membranes, respectively.

Experiments with raw and pretreated produced water, containing 57,000 mg L^{-1} TDS, were carried out in the integrated FO-MD system. Sodium chloride in de-ionized water was used as FO draw solution and was continuously regenerated using the MD system. Although the individual FO process had relatively high rejection to TDS and total organic carbon (TOC), these contaminants could also accumulate in draw solution after a long-term experiment. The MD process provided an effective rejection to contaminants in the draw solution after FO process and the integrated system revealed nearly complete rejection of TDS and TOC. Feed water was concentrated up to 50% in the hybrid system. Next, the concentrated feed was treated using an individual MD and was further concentrated up to 300,000 mg L^{-1} TDS under variety of operating conditions. The results of this study highlight the potential for employing FO-MD to treat high salinity wastewaters from unconventional gas extraction.

533547 Optimizing Sand Control in Wellhead with Advanced De-Sander Technology

Abstract:

Optimizing Sand Control in Wellhead with Advanced De-sander Technology

Banchao Shu, Oluwatosin Oyelakin, Priyanka Shahi, Isaac Snyder, Mike Fredrick, Jeevan Dahal, Certified Holdings LLC- Certified Technologies

Abstract

The well testing equipment usually suffers from severe erosion caused by sand production in conventional and unconventional oil and gas wells. The choke manifold, gate valves, and two-phase/three-phase separators are required to repair and clean up frequently, otherwise, it will cause equipment failure and operation problems. Tremendous time have spent on shut-in wells and repair equipment also.

It is critical to remove the sand from oil and gas production stream and ensure continues operation. The wellhead de-sander can remove sand to protect all the downstream equipment, which enables the easier design and operation of solid-handling systems. This novel multiphase cyclone geometry for wellhead was optimized with Computational fluid dynamics (CFD), and Finite Element Analysis (FEA) methods are presented in this paper. An integrated control system was designed with Ladder Logic for automated and continuous operation.

Methodology

The multiphase cyclonic de-sander geometry was designed and optimized by CFD software ANSYS 18.2. The Volume of Fluid (VOF) model, with the kε-RNG method, was used for multiphase flow simulation.

The nonlinear FEA was used to determine the potential failure locations in the pressure vessels. The failure point was identified in the vessel where the total stress, strain, and elongation exceeds the actual limit of the material. The control system design was designed in Ladder Logic.

Results

A 5000-psi rating sand separation system is designed using Boiler and Pressure Vessel Code (BPVC) Section VIII-Rules for Construction of Pressure Vessels Division 2-Alternative Rules. A cyclonic housing and two sand accumulator are be designed based on this code (Figure 1).

Figure 1 Full diagram of the de-sander system

The separator geometry dimensions and separation efficient can be calculated based on stocks' law, from the following equations. In this model, the multiphase flow spins through a number *N* of revolutions in the outer vortex. The value of *N* can be approximated as the sum of revolutions inside the body and inside the cone (*Equation 1*):

$N = (1/H)\ (Lb + Lc/2)$ *Equation 1*

where

N = number of turns inside the device (no units)

H = height of inlet duct (m or ft)

Lb = length (vertical) of cyclone cone (m or ft).
Lc = length of cyclone body (m or ft)

Lapple developed a semi-empirical relationship to calculate a "50% cut diameter" d_{pc}, which is the diameter of particles collected with 50% efficiency. The expression is (*Equation 2*)

$d_{pc} = 9\mu W/(2*N*V_i*(\rho_p - \rho_a)$ Equation 2

where

d_{pc}= diameter of particle collected with 50% efficiency.

W = width of inlet (m or ft)

V_i = multiphase flow inlet velocity (m/s or ft/s) = Q/WH

ρ_p = density of the particle (kg/m^3)

ρ_a = air/fluid density (kg/m^3)

μ = air viscosity (kg/m.s).

Theodore and DePaola then fitted an algebraic equation to the curve, which makes Lapple's approach more precise and more convenient for application to computers. The efficiency of collection of any size of particle is given by (*Equation 3*)

$= 1/(1+(d_{pc}/d_{pj})^{\wedge}2$ *Equation 3*

where

$_j$ = collection efficiency of particles in the *j*th size range ($0 < j < 1$)

d_{pc} = diameter of particle collected with 50% efficiency

d_{pj} = characteristic diameter of the *j*th particle size range (in μm).

The overall efficiency, called performance, of the cyclone is a weighted average of the collection efficiencies for the various size ranges, namely (*Equation 4*)

$= \Sigma_j\ m_j/M$ Equation 4

where

= overall collection efficiency (0 << 1)

m_j = mass of particles in the *j*th size range

M = total mass of particles.

The cyclonic insert geometry was then optimized by ANSYS 18.2 with multiphase flow simulation based on critics such as higher sand separation efficiency and lower pressure drop. The cyclone separation system can treat up to 100 MMSCFD of natural gas and up to 13,000 bbl/d of liquid with 1-2% of sand. The multiphase simulation result suggests more than 99% sand remove efficiency for a particle size larger than 5 μm.

The FEA analysis was further adapted to determine the failure locations in the pressure vessels. Using the FEA analyses to optimize the shape and the wall thickness of the pressure vessel thereby have prevented failure and reduced the cost of over-engineering in the designed.

Figure 2 FEA analysis for the sand Accumulator

The FEA analysis result suggest the potential failure point in the two-inch water inlet hole (Figure *2*). We then have added fillet to reduce the stress result from the sharpen are.

The control system would allow for a maximum 75 psi pressure drop in the de-sander system to ensure a longer product lifetime. The advanced automation control system was implemented in the designed thereby elimination possible human interference of the whole assembly operation.

Conclusion

The optimized wellhead cyclonic de-sander with CFD, FEA technology will be able to separate the gas, liquid, and sand with the highest efficiency of 99 percent which has never been approached before. It will provide great protection for all the downstream equipment and reduce the maintenance cost. The automated and continuous process will reduce the shutdown time, reduce human error, and improve the overall production efficiency.

AUTHOR INDEX

Abdelsayed, Victor 104
Abedin, Ashraf 81, 93
Able, Chad 123
Adams II, Thomas A. 62
Adesoji, Adesina 95
Adhikari, Birendra 102
Adjiman, Claire S. 16
Ahammad, Monir 178
Ahmed, Shakeel 101
Al Otaibi, Raja 95
Albenze, Erik 20
Alenazey, Feraih 95
Al-Ghafri, Saif Zs. 31
Alkhatib, Ismail I. 15
Alleman, David 124
Allen, Meredith 166
AlMomani, Fares 47
Almutairi, Ghzzai 95
Alotaibi, Bandar 95
Alotaibi, Faisal 95
Alves, Rita M. B. 143
Alvin, Mary Anne 113
Alyousef, Yousef 95
Alzobaidi, Shehab 137
Ampomah, William 24, 28, 29
Anderson, Eric 147
Armstrong, Robert C. 87
Arthur, J. Daniel 124
Assary, Rajeev 167
Austin, Danielle 70
Auyeung, Nick 74
Awati, Rohan 3
Azenkeng, Alexander 134
Bae, Joongmyeon 97, 99
Bae, Minseok 97
Bai, Xinwei 108
Balch, Robert 24
Baltrusaitis, Jonas 77
Banu, Aliya 47
Barmak, Katayun 43
Baroi, Chinmoy 83
Baser, Deven 79
Bazant, Martin Z. 174
Beckham, Gregg T. 147
Bell, Alexis T. 130
Bell, David A. 63
Benson, Steve 117
Berry, David 100
Bhan, Aditya 80
Bhattacharyya, Debangsu 10, 11, 64, 107
Bhattar, Srikar 81, 93
Bhosale, Rahul 47
Biegler, Lorenz T. 88
Biesheuvel, P. M. 174
Blinn, Kevin 41
Boardman, Richard 63
Bock, Sebastian 89
Boldor, Dorin 36, 112
Bonita, Yolanda 165
Brickett, Lynn 9, 40
Buchanan, J. Scott 148
Bunger, Andrew P. 138, 163
Burgard, Anthony P. 64
Burgess, Ward A. 116
Butler, Shane 134
Butterworth, Tom 42
Call, Ann V. 42
Camaioni, Donald M. 141
Cather, Martha 24
Chada, Joseph P. 148
Chadderdon, David 170
Chadderdon, Xiaotong 170
Chen, Bing-Hung 125
Chen, Yuanxin 1
Cheng, Cheng 163
Cheng, Yan 142
Cheng, Yu-Chieh 82
Cheng, Zhuo 79
Cheon, Hyungjun 97
Chhabra, Pulkit 175
Choi, Jae-Wook 75
Christeson, Tyler 91
Ciferno, Jared 133
Clelland, Kate 91
Coblyn, Matthew Young 74
Corgnale, Claudio 102
Crandall, Dustin 139
Culp, Jeffery 139
Curtiss, Larry 167
Da, Chang 137
Dahal, Jeevan 180
Dai, Xiong 27
Damjanac, Branko 138
Danghyan, V. 94
Datye, Abhaya K. 142
Deng, Yifan 82
Desai, Pratik 42
Dewitt, Stephen J. A. 3
Diao, Weijian 102
Dipietro, Phil 59
Doyle, Patrick S. 173
Duan, Pu 168
Dumesic, James A. 148, 167
Eagan, Nathaniel 148
Edwards, Peter P. 98
Emmert, Marion 168
Enick, Robert 137
Epelle, Emmanuel 177
Esser, Richard 21
Eylands, Kurt 134

AUTHOR INDEX

Fan, Liang-Shih 79, 90, 91
Figueroa, J. D. 9
Finn, Matthew 169
Finzel, Jordan 168
Flannery, Matt 37
Frederick, Brian G. 166
Fredrick, Mike 180
Free, Michael 118
Freeman, Charles J. 10
Frimpong, Reynolds A. 12
Fu, Wei 138
Fushimi, Rebecca 83
Fyfe, Peter 179
Galindo, Amparo 16
Gardezi, Syed Ali Zeeshan 32
Garlapalli, Ravinder 136
Garrett, Bruce 19
Gerogiorgis, Dimitrios I. 177
Ghoniem, Ahmed F. 103
Gilbert, William J. R. 135
Ginosar, Daniel M. 102
Giudici, Reinaldo 143
Glezakou, Vassiliki-Alexandra 2
Gohndrone, Thomas 44
Goodman, Angela 139
Goulas, Konstantinos A. 127
Goyal, Amit 46
Grabow, Lars C. 127
Graham, Brendan F. 31
Gramlich, William M. 166
Granados-Focil, Sergio 168
Granite, Evan J. 116, 120, 121
Grant, Tim 25
Griffin, Michael B. 140
Grigg, Reid 24
Grunlan, Jaime C. 5
Guo, Mengqing 90
Guo, Mond 126
Guo, Xinwen 34
Gupta, Vijay 35
Gutierrez, Oliver 141
Ha, Jeong-Myeong 75
Hacker, Viktor 89
Hakala, Alexandra 122
Ham, Hyung Chul 75
Hamling, John A. 22
Han, Yang 1, 6
Harale, Aadesh X. 99, 101
Haribal, Vasudev Pralhad 84
Hariharan, Subrahmaniam 4
Harris, Justin 137
Hartzler, Daniel 164
Hasan, A. Rashid 178
Hatton, T. Alan 4, 174
Haynes, Daniel J. 93, 100
He, Fan 174
Heath, Jason 29
Heldebrant, David J. 2
Heltzel, Jacob 169
Hewer, Thiago L. R. 143
Hicks, Jason C. 165
Ho, W. S. Winston 1, 6
Holmes, Thomas 42
Honaker, Rick 119
Hopkinson, David 14
Howard, Bret H. 116
Hower, James C. 114
Hsu, Emily 43
Hsu-Kim, Heileen 115
Hu, Jianli 107, 108
Huber, George W. 148, 167
Hughes, Ryan 11
Irons, Trevor 21
Jackson, George 16
Jain, Jinesh 164
Jana, Amiya Kumar 38, 39
Jessen, Kristian 23
Jia, Wei 21, 28, 29
Jiang, Xiao 34, 45
Jie, Xiangyu 98
Jin, Xin 131
Jitsukawa, Koichiro 132
Johnson, Robert L. 142
Johnson, Stephen J. 135
Johnston, Keith 137
Jovanovic, Goran 74
Kalogerakis, Nicolas 173
Kanitkar, Swarom 81, 93
Kapellos, George E. 173
Karimi, Iftekhar A. 175
Katahira, Rui 147
Kathe, Mandar 91
Katikaneni, Sai P. 97, 99, 101
Kauffman, Douglas R. 106
Keller, Murphy J. 116
Khraisheh, Majeda 47
Kiani, Daniyal 77
Kim, Jin-Kuk 8
Kim, Jong Suk 63
Kim, Seonah 140
Klapperich, Ryan J. 22
Koech, Phillip K. 2
Kong, Frank 91
Kopp, Daniel 41
Kotamreddy, Goutham 11
Kraft, Markus 175
Krishna, Siddarth H. 167
Kumar, Anand 47
Kumar, Anikesh 175
Kumar, Anurag 80

AUTHOR INDEX

Kurz, Bethany 134
Kutchko, Barbara 139
Kweon, Hyukmin 30
Kwon, Joseph Sangil 162, 178
Ladwig, Ken 120
Lai, Haoxiang 62
Lai, Yungchieh 82
Lail, Marty 35, 44
Laudal, Dan 117
Lee, Andrew 64
Lee, In-Beum 172
Lee, Jaemyung 99
Lee, Jason J. 137
Lee, Sunghoon 8
Lekse, Jonathan W. 116
Lercher, Johannes A. 141
Li, Fanxing 84
Li, Wenzhen 170
Li, Yongdan 33
Li, Zhong 26, 27
Liang, Jenn-Tai 135
Lim, Seo Yeon 75
Lima, Rubens W. S. 143
Lin, Eric 146
Lin, Ronghong 121
Lin, Shawn D. 171
Litynski, John 40
Liu, Dongxia 78
Liu, Kunlei 12, 60
Liu, Sibao 128
Lively, Ryan 3
Llovell, F. 15
Long, Jeffrey R. 10, 11
Lu, Congwen 137
Luterbacher, Jeremy S. 129
Lynn, Bert C. 36
Ma, Jinliang 60
Macala, Megan 14
Mahajan, Kanwal 9, 20
Mahapatra, Priyadarshi 60
Malhotra, Deepika 2
Mannan, M. Sam 178
Mantripragada, Hari C. 92
Marculescu, Cosmin 112
Marin, Christopher 106
Martell, Spencer 166
Massen-Hane, Michael 4
Matuszewski, Michael S. 10, 11, 60
May, Eric F. 31
McIntyre, Dustin 164
McPherson, Brian 21, 28, 29, 30
Mevawala, Chirag 107
Miao, Guang 26, 27
Miao, Yu 74
Mibeck, Blaise 134
Miller, David C. 10, 64
Mitchell, Scott F. 67
Mitsudome, Takato 132
Mizugaki, Tomoo 132
Moate, Joseph 59
Mobley, Paul 35
Mohamed, Omar 74
Monnier, John 102
Moodie, Nathan 21, 28, 29
Morgan, David 25
Mukarakate, Calvin 140
Mukasyan, A. 94
Muley, Pranjali 36
Naceur, Kamel Ben 18
Nadgouda, Sourabh 79, 90
Nakles, David 22
Natesakhawat, Sittichai 139
Neal, Luke 84
Neumayer, Tony 59
Nguyen, Manh 2
Nie, Xiaowa 34
Nikolic, Heather 12
Nimlos, Mark 140
Nitta, Kodai 132
Novoa, S. Calderon 94
Obilade, Olatoyosi 149
O'Connell, Timothy 165
Ogden, David 123
Oh, Jiwoo 97
Oh, Su Cheun 78
Okoli, Chinedu O. 64
Omell, Benjamin P. 10, 60
Ostace, Anca 64
Oyelakin, Oluwatosin 180
Palodkar, Avinash V. 38
Papadokonstantakis, Stavros 16
Papadopoulos, Athanasios I. 16, 57
Papadourakis, Antonis 69
Park, Ah-Hyung Alissa 43
Pasel, Joachim 96
Patil, Vivek 30
Patrascu, Mariana 112
Pekney, Natalie J. 61
Pelgen, Jonathan V. 12, 60
Pennisi, Kenneth J. 7
Perdomo-Hurtado, Felipe 16
Pereira, Luis M. C. 15
Perry, Robert J. 137
Peters, Ralf 96
Pham, Hien N. 142
Powell, Joseph B. 66
Pyka, Anthony 74
Qin, Lang 79
Rahimi, Mohammad 4
Rahman, Fahim 102

AUTHOR INDEX

Ramasamy, Karthikeyan K. 126
Rashke, Quinn A. 167
Rayer, Aravind V. 35
Realff, Matthew 3
Reddick, Ian 74
Reed, Michelle 147
Remson, Donald 25
Resnik, Kevin P. 14
Rew, Brittany 117
Riman, Richard E. 41
Robinson, Brandon 108
Rodosta, Traci 20
Rodriguez, Gianfranco 137
Rokkam, Srujan 37
Rollins, Harry W. 83
Roman-Leshkov, Yuriy 147
Rorrer, Julie 130
Roth, Elliot 116, 120
Rothman, Rachael H. 42
Rousseau, Roger 2
Rozmyslowicz, Bartosz 129
Ryu, Jun-Hyung 172
Saha, Basudeb 128
Sahinidis, Nick 176
Salim, Witopo 1
Samad, Jadid E. 46
Samsun, Remzi Can 96
Sandefur, Evan 169
Sanguinito, Sean 139
Sardari, Kamyar 179
Sauk, Benjamin 176
Savitski, Alexei A. 138
Schaffers, William C. 63
Schmal, Martin 143
Schmidt, Zachary R. 167
Schmidt-Rohr, Klaus 168
Schwartz, Thomas J. 166
Scurto, Aaron M. 135
Seferlis, Panos 16, 57
Sengupta, Arupananda 71, 109
Shafiei, Mohammad 76
Shahi, Priyanka 180
Shakib, Akbar Mahdavi 166
Shangguan, Ning 7
Shanks, Brent H. 142
Shareghi, Adam 74
Sharma, Lohit 77
Shavalieva, Gulnara 16
Shekhawat, Dushyant 93, 100, 104
Shi, Jian 36
Shi, Wei 14
Shi, Zhuofan 23
Shih, Yi-Chen 125
Sholl, David S. 3
Shu, Banchao 180
Siahvashi, Arman 31
Siddhamshetty, Prashanth 162, 178
Siefert, Nicholas 14, 122
Siegelman, Rebecca 10, 11
Siirola, Jeffrey J. 65
Simonetti, Dante 146
Singh, Madhu 71, 109
Skoptsov, George 71, 105, 109
Smith, Mark W. 104
Smith, Ryan 115
Snyder, Isaac 180
Song, Chunshan 34, 45
Song, Hua 70
Song, Kwang Ho 75
Song, Yuying 127
Soong, Yee 121
Soukri, Mustapha 35
Sourav, Sagar 77
Spencer, Michael 136
Spivey, Jim 81, 93, 100
State, Razvan N. 112
Stolaroff, Joshuah 10
Stolten, Detlef 96
Stone, Michael 147
Suh, Dong Jin 75
Sullivan, Mary 20
Sun, Lin 23
Sun, Yuhan 86
Taggart, Ross 115
Takalkar, Gorakshnath 47
Tan, Eric C. D. 140
Tanthana, Jak 35
Tengco, John Meynard M. 102
Thakre, Niraj 39
Theaker, Nolan 117
Thompson, Robert L. 14
Thompson, Samuel 44
Timko, Michael T. 168
Tiwari, Sarojini 108
Tiwari, Surya 14
Tong, Andrew 13, 91
Toste, Dean 127, 130
Traverso, Andrew 74
Trembly, Jason 123, 136
Trujillo, Edward 30
Truong, Quang 37
Tsau, Jyun Syung 135
Tschauder, Andreas 96
Tse, Lik Hang Hugo 42
Tsivintzelis, Ioannis 57
Tsotsis, Theodore 23
Tyufekchiev, Maksim 168
Vakharia, Varun 1
VanDer Wal, Randy 71, 105, 109
Vega, Lourdes F. 15

AUTHOR INDEX

Vergara, A. J. C. 61
Veser, G. 82, 92
Vlachos, Dionisios G. 128
Vo, Dai-Viet N. 95
Voutchkova-Kostal, Adelina 169
Wachs, Israel E. 77
Walton, Krista S. 34
Wang, Aiguo 70
Wang, Jun 41
Wang, Kai 169
Wang, Meng 141
Wang, Miao 4
Wang, Wenjia 45
Wang, Xiaoxing 45
Wang, Ying 63
Wenzlick, Madison 122
West, Alan C. 43
West, David 68
Wickramasinghe, S. Ranil 179
Wiesner, Mark R. 115
Wildfire, Christina 104
Wilhite, Benjamin 5
Will, Robert 24
Winberg, Steven 85
Wittrig, Ashley 148
Wojnar Jr., Theodore J. 17
Wolf, Eduardo E. 94
Womble, Charles 59
Wu, Dongzhu 1
Wu, Xiao-Yu 103
Xiao, Chongwei 149
Xiao, Jing 26, 27
Xiao, Tiancun 98
Xiao, Ting 21, 28
Xu, Boyue 178
Xu, Zhijie 10
Yan, Hao 131
Yang, Chaohe 131
Yang, Cuiting 26
Yang, Sheng-Chiang 171
Yang, Yahui 82
Yeap, Jher Hau 129
Yin, Bin 131
Yokochi, Alexandre 74
Yoo, Jae Young 99
Yu, Jiah 172
Yun, Seokwon 8
Yung, Matthew M. 140
Zacharias, Robert 89
Zarogiannis, Theodoros 57
Zeller, Kurt 71, 105, 109
Zhang, Guangyu 131
Zhang, Yitao 13
Zheng, Feng 2
Zheng, Qinghe 44
Zhou, S. James 35
Zhu, Yizu 76
Zimmerman, William B. 42
Zwoster, Andy J. 2

AIChE
120 Wall Street, FL 23
New York, NY 10005-4020

ISBN 978-1-5108-7642-2

9 781510 876422

CURRAN ASSOCIATES INC.
proceedings.com